Solar Drying Systems

T0176051

Solar Drying Systems

Om Prakash

Anil Kumar

CRC Press
Taylor & Francis Group
Boca Raton London New York

CRC Press is an imprint of the
Taylor & Francis Group, an **informa** business

First edition published 2020
by CRC Press
6000 Broken Sound Parkway NW, Suite 300
Boca Raton, FL 33487-2742

and by CRC Press
2 Park Square, Milton Park, Abingdon, Oxon, OX14 4RN

© 2021 Taylor & Francis Group, LLC

CRC Press is an imprint of Taylor & Francis Group, LLC

Library of Congress Cataloging-in-Publication Data

ISBN: 978-0-367-28043-7 (hbk)
ISBN: 978-0-429-29935-3 (ebk)

Typeset in Palatino
by SPi Global, India

Contents

Foreword by Rebecca R. Milczarek

Much regarding the solar drying of food materials has remained unchanged in the 14,000 years that cultures around the globe have practiced it. Today, millions of entities—ranging from smallholder farmers in developing regions to multinational companies operating on thousands of hectares—still employ open sun drying to preserve fruit, vegetables, grains, and pulses. The continued popularity of open sun drying is because it is perhaps the most widely accessible and low-cost food preservation method available.

While much in this field has remained the same, since the 1990s there have been rapid technological advancements in solar drying that promise to enhance product quality, shorten drying time, and (further) improve the environmental sustainability of the process. While open sun drying still has a strong presence, indirect, mixed-mode, and active solar dryer designs are becoming increasingly popular. A deeper understanding of solar drying's mass and energy transport phenomena is now possible through computational multiphysics modeling programs. New dryer designs can be disseminated across the globe in mere seconds. The drive to reduce greenhouse gas and carbon emissions in the food processing industry has also renewed interest in exploring and optimizing solar drying technologies.

Solar Drying Systems is a text that meets this moment. Drs Prakash and Kumar have brought their collective 15-plus years of solar drying research and teaching experience to bear on this comprehensive reference and instructional work. The authors have studied the solar drying of a diverse array of crops and food products—apple, bitter gourd, gooseberry, jaggery, onion, tomato, and watermelon, to name a few. The Agricultural Engineering community widely cites their physics-driven approach to designing and analyzing solar drying systems' performance, and this approach forms the framework of their book. Readers will gain an understanding of both the fundamental principles of solar drying, which have remained constant over millennia, as well as recent advances in dryer design, analysis, and optimization, which are enabling a modern reimagining of the humble solar dryer.

Looking forward to a bright future,
Dr. Rebecca R. Milczarek
Research Agricultural Engineer
September 2020

Preface

As the rise in demand for fossil fuel has been increasing on day-to-day basis there is a need to put more attention on renewable energy sources. The sun is always being considered as an unlimited source of energy. The use of solar energy is not a new concept; it has been used since earlier in different ways. Solar drying is one of the most prominent applications of the solar energy. It is being used since long time for drying of different agricultural products for farmers and small scale agro based industries.

There is a rapid advancement in the field of solar dryer due to intensive research in the field of solar drying. Due to consistent effort from last three-decade, solar drying system becomes cost effective as well as energy efficient.

This book presents complete information related to solar dryer. The whole book is divided into seven chapters.

The first chapter deals with various fundamentals related to the solar drying system. The solar drying system is mainly used to dry the agricultural produce in the low thermal application. It is good in three prospective mainly economy, environment, and minimization of post harvest loss. In this chapter, importance of drying is also discussed in details Various others related topics also discussed like crop drying characteristics, safe storage moisture contents, dried product quality parameters and classification of drying. At last this chapter, put foundation of the solar drying.

The second chapter deals with various drying methodology for the solar dryer. This methodology is required to make drying process effective. Various related concepts are to be discussed in detail such as moisture content, equilibrium moisture content, drying rate, shrinkage and pre-treatment prior to drying. By proper understanding the concept, it becomes easy to take maximum advantage of the drying process.

In the third chapter, a compressive review is being presented of the various state of art solar dryer. Various type of solar dryer with different design are discussed in this chapter.

Chapter four deals with the performance analysis of the solar drying system. This analysis is very important to justify the utility of the solar dryer. Any newly developed solar dryer is being evaluated based on these performance parameters and compared the results with previous dryer. The drying efficiency, heat utilization factor, coefficient of performance, thermal efficiency, overall daily thermal efficiency, and exergy analysis are the prominent performance analysis parameters based on that any solar dryer is being judged. These performance parameters are applicable in the all the dryers with slight modification, which vary from case to case to basis.

Chapter five deals with the thermal modelling of the solar dryer. By the help of thermal modelling, these parameters can be predicted with high accuracy namely inside air temperature, inside air relative humidity drying rate, drying kinetics, and drying potentials.

Chapter 6 present the energy analysis of the solar dryer. All the important parameters related to energy analysis such as embodied energy, energy payback time, carbon credit and CO_2 mitigation are being discussed.

The last chapter deals with economic analysis of the solar dryer. All important economic parameters are discussed in this such annual cost, payback period and others.

We hope that the present content of the book in respect of different technologies of solar drying can serve the useful information to practicing engineers, learner's, faculty members, and various student of the world. Despite our best concern and effort, we apologize if some errors are in the manuscript due to inadvertent mistakes. We would appreciate being informed about the errors and constructive criticism for the enhancement of the quality of this book.

Om Prakash
Ranchi, India

Anil Kumar
New Delhi, India

Acknowledgments

This book is a tribute to the engineers and scientists who continue to push forward the practices and technologies of solar drying systems. It would not have been completed without the efforts of numerous individuals in addition to the primary writers, contributing authors, technical reviewers, and practitioners.

Our first and foremost gratitude is that we have been given the opportunity and strength to play our part in the service of society. We also express our heartfelt gratitude to Prof. Yogesh Singh, Vice Chancellor, Delhi Technological University, Delhi, India, and the Vice Chancellor of Birla Institute of Technology, Mesra, Ranchi, India, for their kind encouragement.

We would like to express our thanks to many people, faculty and friends who provided valuable inputs during the preparation of this text. We thank especially Dr. Jan Banout, Czech University of Life Sciences, Prague, Czech Republic; Prof. Ahmed M. Abdel-Ghany, King Saud University, Saudi Arabia; Prof. Shuli Liu, Coventry University, UK; Prof. Cristina L. M. Silva, Universidade Católica Portuguesa, Portugal; Prof. Raquel P. F. Guiné, Polytechnic Institute of Viseu, Portugal; Dr. Ashish Shukla, Coventry University, UK; Prof. Perapong Tekasakul, Prince of Songkla University, Hat Yai, Songkhla, Thailand; Prof. Samsher, Delhi Technological University, Delhi (India); Prof. (Retd.) G. N. Tiwari, Centre for Energy Studies, Indian Institute of Technology, Delhi, India; and Prof. Emran Khan, Head of Department of Mechanical Engineering, Jamia Millia Islamia, New Delhi, India, for their valuable help and suggestions.

We wish to acknowledge our research scholar, Mr. Asim Ahmad, Mechanical Engineering Department, Birla Institute of Technology, Mesra, Ranchi, India, and all colleagues for their support and encouragement.

We appreciate our spouses, Mrs. Poonam Pandey and Mrs. Abhilasha, and our beloved children Ms Shravani Pandey, Master Tijil Kumar, and Ms. Idika Kumar. They have been a great source of support and inspiration, and their endurance and sympathy throughout this project have been most valued. Our heartfelt special thanks go to CRC Press, for publishing this book. We would also like to thank all those who have been involved, directly or indirectly, in bringing the book to fruition.

Finally, yet importantly, we wish to express our warmest gratitude to our respected parents – Sh. Krishna Nandan Pandey, Smt. Indu Devi, the late Sh. Tara Chand, Smt. Vimlesh, and our siblings for their unselfish efforts to help in all fields of life.

Authors

Dr. Om Prakash is working as assistant professor in the Department of Mechanical Engineering, Birla Institute of Technology, Mesra, Ranchi, India. This university is the premier technical university in India, established in 1955 by Mr. B. M. Birla as a technical institute and deemed equal to a university. Main campus size: 780 acres; total number of registered students across all centers: more than 10,000; doctoral students: 300; programs offered: undergraduate, post-graduate, and doctoral.

Dr. Prakash's experience is in teaching and research. He has nine years' experience in the field of energy technology. His areas of specialization are: energy technology, renewable energy, solar energy applications, energy economics, heat transfer, daylighting, energy conservation, and biomedical. He has one government-funded project of 18.5 lakhs.

He has published 30 papers in international peer-reviewed journals, 15 papers in international/national conference proceedings, and 16 chapters in books published internationally. His papers are cited in reputed relevant journals. He has received more than 892 citations, with an h-index of 17. With his supervisor, he developed a thin-layer drying model in 2014, known as the "Prakash and Kumar model." This model is used and cited by many researchers around the globe. He is the author of four books (one national and three international editions). He is a contributing author to a book published in 2020 by CRC Press, *Energy Management: Conservation and Audits*. He has also filed and published two patents—one for a solar dryer and another in biomedical. He is supervising two Ph.D. scholars and 10 Masters students.

Dr. Anil Kumar is associate professor in the Department of Mechanical Engineering, with additional charge of Additional Coordinator, Centre for Energy and Environment, Delhi Technological University, Delhi, India. He gained his Ph.D. in Solar Energy from the Centre for Energy Studies, Indian Institute of Technology, Delhi, India, in 2007. He was a post-doctoral researcher at the Energy Technology Research Center, Department of Mechanical Engineering, Faculty of Engineering, Prince of Songkla University, Hat Yai, Songkhla, Thailand, in the discipline of Energy Technology from June 2015 to May 2017. He also served as assistant professor at the Energy Centre, Maulana Azad National Institute of Technology Bhopal, India, from 2010 to 2018, and assistant professor in the Department of Mechanical Engineering, University Institute of Technology, Rajiv Gandhi Proudyogiki Vishwavidyalaya, Bhopal, India, from 2005 to 2010. The nature of his experience is teaching and research (science, technology, society, and sustainable development). His areas of specialization are: energy technology, renewable energy, solar energy applications, energy economics, heat transfer, natural rubber sheet drying, and environmental issues, and has successfully completed many research funded projects in these areas. He has more than 14 years' experience in the field of energy technology. He has published 120 papers in international peer-reviewed journals and 75 papers in the international/national conference proceedings. His papers are cited in all the reputed relevant journals. He has received more than 3000+ citations, with a 30 h-index. Working with his student, he developed a thin-layer drying model in 2014; it is known as the "Prakash and Kumar model." This model is used and cited by many researchers around the globe. He is an author of nine books (four national and five international editions). He is a contributing author to the book entitled *Energy Management: Conservation and Audits*, published by CRC Press in 2020.

He has also filed and published three patents—one on solar dryers and another on solar photovoltaic thermal collectors. He has supervised seven Ph.D. scholars and 33 Masters students. Dr. Kumar has visited a number of countries, including the UK, Thailand, and Malaysia, and has received various awards and appreciation from reputed sources. Some of these are listed below:

- *Commendable Research Award 2019*: for publishing quality research papers in Delhi Technological University, Delhi, India.

- *Research Excellence Award 2016*: the researcher has Top 20 publications from the Web of Science database, honored by Dr Chusak Limsakul, President Prince of Songkla University, Hat Yai, Thailand.

- *Appreciation for outstanding services* from Chairman BOG and Director, M.A.N.I.T., Bhopal, India, International Institute of Engineers, Kolkata, Elsevier, etc.

- *Best Paper Award in Global Conference on Energy and Sustainable Development (GCESD 2015)*, February 24–26, 2015 at Coventry University Technology Park, Puma Way, Coventry, West Midlands, UK.

1

Fundamentals of Solar Drying Systems

1.1 Introduction

Plants are the primary source of food for the human population. Human food is further categorized as perishable, non-perishable, processed, raw, fresh, manufactured, organic and foods usable in various ways. The preservation of food is also crucial. The Food and Agriculture Organization of the United Nations (FAO) reported that many people in developing countries are acutely malnourished (www.fao.org), and it is certain that extra crops will need to be cultivated to guarantee food safety. This situation is mainly a result of an increase in population (Chauhan and Kumar, 2016). An alternative solution is to minimize loss at the levels of pre-harvest, during harvest, and post-harvest. Among these three types of loss of the crop, post-harvest loss is the most prominent. Post-harvest loss occurs mainly because of microbial degradation of the crop. Microbial growth depends primarily on storage conditions and moisture content. However, different kinds of microorganisms have dissimilar growth rates, relating to conditions such as product, storage atmosphere, content, etc. (Matthews et al., 2019).

The main methods of preservation to ensure the minimization of post-harvest loss of the food product are freezing, vacuum sealing, canning, and preserving with sugar, irradiating fruit, applying preservatives, and drying.

Drying is the most environomical method for minimization of post-harvest loss. The process can be used for various types of food products for long-term preservation with minimum compromise to the product quality, texture, and color (Aumporn et al., 2018). The drying process is a very energy-intensive process. Figure 1.1 represents energy consumption for drying as seen from the global perspective.

Dried food takes up very little space compared to frozen and canned food. However, it cannot be used as a substitute for freezing and canning methods because of their capacity to retain taste, appearance and nutritive value. Drying occurs through heat and mass transfer of the dryable product. The flow of energy is presented in Figure 1.2.

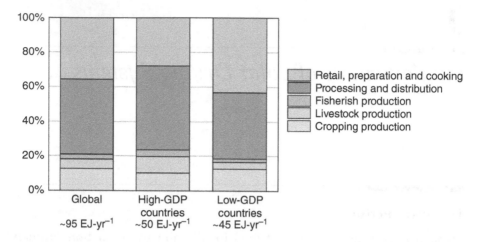

FIGURE 1.1
Global energy use in the food sector (Lamidi et al., 2019).

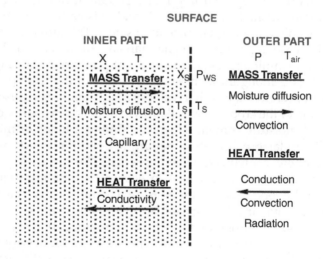

FIGURE 1.2
Transportation of heat and mass transfer from the inner to the outer part of the product.

There has been intensive research in the field of drying, with the aim of making it more effective and less expensive. Every year, more and more publications are produced regarding this field. A graphical representation is shown in Figure 1.3.

To take maximum advantage of drying, it is very important to select pre- and post-processing of the product both before and after the drying event.

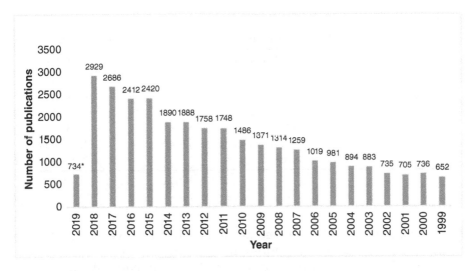

FIGURE 1.3
Trend of research publications from 2000–2019 (Menon et al., 2020).

The selection of a suitable dryer is also an important consideration in minimizing the drying time and retaining the nutrient content of the dried product. Based on the capacity of the dryer, the drying load per batch is to be estimated. The pre-processing of the dryable products includes osmotic dehydration, blanching, salting, and soaking. However, based on the product to be dried, the pre-processing method varies from case to case. The main post-processing operations carried out on the dried products are coating and packaging. These have a deep impact on preserving the product over a long time span.

1.2 Importance of Drying

It is well known that developing countries are facing problems of food security. There are two ways to attain this food security. The first is to increase productivity by more farming, and the second is to minimize losses. There are serious problems in increasing farming because it is already in the saturation stage. In India, agricultural commodities contribute 18% of the gross domestic product (GDP) of the country. Agriculture also provides 50% of the employment of its citizens. Hence there is very little scope to increase the area of agriculture. However, recent studies reveal that, on average, 16% of

TABLE 1.1

Post-Harvest Losses of Various Prominent Agricultural Commodities in India

Serial No.	Commodities	% Loss
1	Grains	4.65–5.99
2	Pulses	6.36–8.41
3	Fruits and vegetables	4.58–15.88
4	Floriculture	30–35
5	Medicinal and aromatic plants	Not estimated
6	Oil seeds	3.08–9.96

the agricultural produce is wasted at the post-harvest level. Crop can also be lost in the pre-harvest and during harvest, but post-harvest loss is very prominent. Table 1.1 shows the post-harvest loss of agricultural commodities in India.

It is therefore very important to minimize this loss at the post-harvest level with a minimum investment of both energy and economy.

Drying emerges as one of the most prominent solutions to minimizing post-harvest loss with the least use of energy and economy. It is a complex method that involves at the same time heat and mass transfer along with several other processes. This results in physical/chemical damage which leads to variations in the quality and shape of the product. These are the prominent changes that occur during the drying process, i.e., shrinkage, crystallization, puffing, transformations from shrinkage, variance in color, texture, and aroma, along with other changes that occur as a result of a compound reaction (Singh and Shrivastava, 2017).

The important features of food drying are as follows:

1. The working temperature has a wide range, from triple point to the critical point of the produce's moisture level.
2. Working pressure is in a wide range, with the lower atmospheric pressure to 25 bar.
3. Drying loads are across a wide range, from 0.10 kg/h to 100kg/h.
4. The drying time has a wide range, from less than 1 second to 5 months.
5. The thickness of the produce is in wide ranges from microns to tens of centimeters.
6. Porosity of the produce is in a wide range, from 0%–99.9%.
7. All three types of heat transfer take place simultaneously.

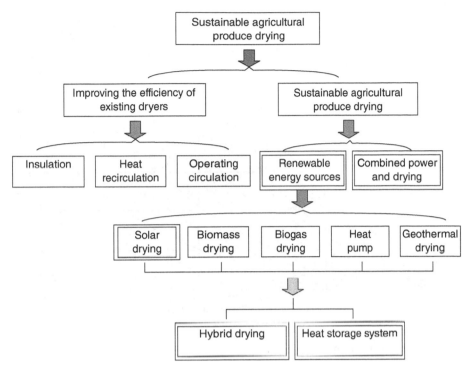

FIGURE 1.4
Classification of sustainable agricultural produce drying (Lamidi et al., 2019).

Drying is a highly energy-consuming process which uses 10–20% of total energy from various developed countries. Hence it is very important to develop energy efficient drying. Another way of saving energy is by using renewable energy sources for the process (Kumar et al., 2017). Figure 1.4 presents the broad classification of the drying of sustainable agricultural produce. This includes both renewable and non-renewable energy-driven dryers.

The behavior of the moist air can be explained clearly with the help of a psychometric chart, and the chart can also explain the drying phenomenon. In drying there are two major parameters, namely temperature and humidity. Both parameters are shown in the chart. Hence researchers and student can use this chart to determine the drying behavior. A sample of a psychometric chart is given in the Figure 1.5.

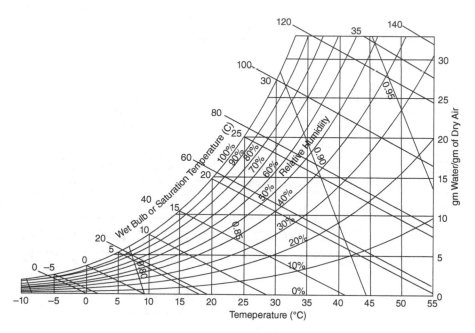

FIGURE 1.5
A classical psychometric chart.

1.3 Crop Drying Characteristics

The main aim of the crop drying process is to reduce the amount of water content in the crop to a safe storage level in order to increase its shelf life (Elkhadraoui et al., 2015). After the product is dehydrated or dried, the rate of degradation will be decreased because of the reduction of microbial activity and resulting biochemical reactions. It eliminates not only post-harvest losses but also transport costs and storage space once a large quantity of water has been removed from the crop. Different crops contain varying amounts of moisture.

Table 1.2 lists selected crops with their initial and final moisture content. To ensure a long shelf life, each crop is dried to reach a different final moisture content. To attain the chosen moisture content, crops have to be dried at the maximum allowable temperature to avoid over- or under-drying. Both of these states have a negative impact on a crop's shelf life, therefore it is essential to maintain the maximum allowable temperature when drying.

Crops can also be classified based on moisture content. There are three types of crops, with high, medium, and low moisture content. High moisture content crops contain moisture greater than 80%, medium moisture content crops contain moisture of between 30% and 80%, and the remaining crops

TABLE 1.2

Initial and Final Moisture Content and Maximum Allowable Temperatures for Drying a Range of Crops

Crop	Initial Moisture Content (%, w.b.)	Final Moisture Content (%, w.b.)	Maximum Allowable Temp. (°C)
Paddy rice, raw	22–24	11	50
Paddy rice, parboiled	30–35	13	50
Maize	35	15	60
Wheat	20	16	45
Corn	24	14	50
Rice	24	11	50
Pulses	20–22	9–10	40–60
Oil seed	20–25	7–9	40–60
Green peas	80	5	65
Cauliflower	80	6	65
Carrot	70	5	75
Green beans	70	5	75
Onion	80	4	55
Garlic	80	4	55
Cabbage	80	4	55
Sweet potato	75	7	75
Chilies	80	5	65
Apricot	85	18	65
Apples	80	24	70
Grapes	80	15–20	70
Bananas	80	15	70
Guavas	80	7	65
Okra	80	20	65
Pineapple	80	10	65
Tomatoes	96	10	60
Brinjal*	95	6	60

Prakash and Kumar (2014).
* Eggplant.

come under the category of low moisture content crops. Each crop has different moisture diffusivity, which is the rate of movement of moisture without any specified mechanism.

Therefore, based on the type of crop, drying parameters must be adjusted in order to reach the best outcome from the drying process. For low moisture content crops, a thick-layer drying process is recommended, but for medium and high moisture content crops, a thin-layer drying process is advised. The relative humidity also plays a very important role in the drying process. The relative humidity of the drying chamber should be kept to a low level.

1.4 Safe Moisture Content

Safe moisture content (SMC) is a very important parameter for dried products, as each product has an individual safe moisture content level. The list in Table 1.2 presents details of some selected crops with initial and final moisture contents. The final moisture content is the same as the safe moisture content of the crop. If the moisture of the dried product is greater than the safe moisture content then the shelf life of the product will be reduced considerably and products will quickly degrade, leading to huge financial consequences. Hence it is very important to keep the moisture content of the dried product within the safe moisture limit.

1.5 Dried Product Quality Parameters

There is a global emphasis on the quality of dried food products because of its important impact on human health. This includes both external and internal variables for product quality (Huddar and Kamoji, 2019).

To ensure a satisfactory level of nutrient content in food products, standardization is being adopted, and both national and international standards have been set up to monitor dried food production. Approved products are usually tagged with a suitable symbol (Agricultural Products Act, 1937, under the Ministry of Agriculture).

There are two major consideration in the case of dried food products: external and internal factors. External factors include shrinkage, color, flavor/taste, texture, etc.. while internal factors are classified into four subdivisions—namely nutritional content, chemical content, thermo-physical, and microbial.

The nutritional content includes loss of proteins and vitamins, and minerals such as potassium, magnesium, etc. Chemical content includes discoloration, browning, etc. The thermo-physical category includes thermal conductivity, specific heat, bulk and actual density, and surface fracture. The microbial subdivision includes primarily contamination by microorganisms.

1.6 Classification of Drying

There are many types of the drying methodology available across human society. Figure 1.6 presents a detailed classification of the drying system.

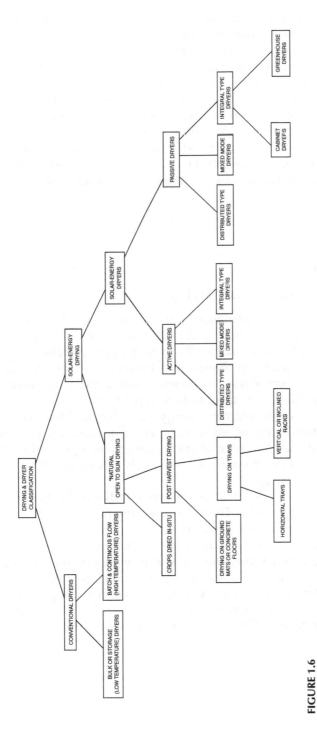

FIGURE 1.6
Classification of drying and dryers (Ekechukwu and Norton, 1999).

The drying process is classified into two types, conventional and solar. Conventional dryers cover all types of mechanical and electrical dryers. They are very harmful to the atmosphere and costly in operation. In contrast, the operating cost of solar dryers is almost zero and it is very environmentally friendly. Solar dryers have been used for generations to dry agricultural produce using open sun drying, also called natural sun drying. But one disadvantage of this method is that, because of changes in the climate, it is not always reliable. Hence controlled solar drying is becoming increasingly popular. Controlled solar dryers are operated in both modes of heat transfer, namely the natural convection mode, and the forced mode. The solar dryer is classified based on the operational mode of heat transfer (Dhalsamant et al., 2018).

1.7 Introduction to Solar Drying

Solar energy is the most prominent source of renewable energy because it is both plentiful and non-polluting. It is also sustainable, affordable, and eco-friendly. The Earth receives some 4000 trillion kWh of electromagnetic radiation every day from the sun. The use of solar energy in tropical climatic zones is very feasible, where at least six hours of sun radiation is received daily at approximately 520–820 W/m² of radiation per hour.

The drying processes are very energy intensive, and the primary energy sources used for drying are fossil fuels. The depletion rate of fossil fuels is very high, and therefore solar energy emerges as a promising alternative energy source. Huge amounts of fossil fuels can be saved by using solar energy (Kumar et al., 2016).

A solar dryer is a system that utilizes solar radiation for the drying operation, as shown in Figure 1.7. There are primarily two types of solar dryer, the open sun dryer and the managed dryer.

Because of changes in the climate across the world, the use of open sun drying is becoming limited, and the use of managed solar drying is more effective in this situation. Such solar dryers can be operated by a forced, natural, or mixed mode of heat transfer (Kant et al., 2016). A proper solar dryer is selected based on location and necessity. The solar drying system can be used either for drying thin layers or for deep bed drying.

FIGURE 1.7
Natural solar drying (Prakash and Kumar, 2014).

Problems

    1.1 Describe the need for a solar drying system in the current scenario?

    1.2 Explain the importance of the drying of agricultural produce.

    1.3 Critically analyze crop drying characteristics.

    1.4 Critically analyze the safe storage moisture content.

    1.5 Define the quality for a dried product and its importance?

    1.6 Classify the various types of drying processes.

    1.7 Discuss the introduction of solar drying in the present scenario.

References

Aumporn, O., Zeghmati, B., Chesneau, X. and Janjai, S., 2018. Numerical study of a solar greenhouse dryer with a phase-change material as an energy storage medium. *Heat Transfer Research*, 49(6), 509–528.

Chauhan, P.S. and Kumar, A., 2016. Performance analysis of greenhouse dryer by using insulated north-wall under natural convection mode. *Energy Reports*, 2, 107–116.

Dhalsamant, K., Tripathy, P.P. and Shrivastava, S.L., 2018. Heat transfer analysis during mixed-mode solar drying of potato cylinders incorporating shrinkage: Numerical simulation and experimental validation. *Food and Bioproducts Processing*, 109, 107–121.

Ekechukwu, O.V. and Norton, B., 1999. Review of solar-energy drying systems II: an overview of solar drying technology. *Energy Conversion and Management*, 40(6), 615–655.

Elkhadraoui, A., Kooli, S., Hamdi, I. and Farhat, A., 2015. Experimental investigation and economic evaluation of a new mixed-mode solar greenhouse dryer for drying of red pepper and grape. *Renewable Energy*, 77, 1–8.

Huddar, V.B. and Kamoji, M.A., 2019, March. Experimental investigation on performance of small passive solar greenhouse dryer for cashew kernel drying. In *AIP Conference Proceedings* (Vol. 2080, No. 1, p. 030001). AIP Publishing.

Kant, K., Shukla, A., Sharma, A., Kumar, A. and Jain, A., 2016. Thermal energy storage based solar drying systems: A review. *Innovative Food Science & Emerging Technologies*, 34, 86–99.

Kumar, A., Deep, H., Prakash, O. and Ekechukwu, O.V., 2017. Advancement in Greenhouse Drying System. In *Solar Drying Technology* (pp. 177–196). Singapore: Springer.

Kumar, M., Sansaniwal, S.K. and Khatak, P., 2016. Progress in solar dryers for drying various commodities. *Renewable and Sustainable Energy Reviews*, 55, 346–360.

Lamidi, R.O., Jiang, L., Pathare, P.B., Wang, Y. and Roskilly, A.P., 2019. Recent advances in sustainable drying of agricultural produce: A review. *Applied Energy*, 233, 367–385.

Matthews, K.R., Kniel, K.E. and Montville, T.J., 2019. *Food Microbiology: An Introduction.* John Wiley & Sons.

Menon, A., Stojceska, V. and Tassou, S., 2020. A systematic review on the recent advances of the energy efficiency improvements in non-conventional food drying technologies. *Trends in Food Science & Technology*, 100(June), 67–76.

Prakash, O. and Kumar, A., 2014. Solar greenhouse drying: A review. *Renewable and Sustainable Energy Reviews*, 29, 905–910.

Singh, P. and Shrivastava, V., 2017. Thermal Performance Assessment of Greenhouse Solar Dryer under Passive Mode. *International Journal of Advanced Technology in Engineering and Science*, 5(5), 530–538.

2

Drying Methodology

2.1 Introduction

In the drying process, moisture removal from a product requires a simultane-ously transfer of mass and heat (Bala and Bala, 1997), which lead to changes in the physical structure and chemical composition of the product (Chauhan and Kumar, 2016). These changes are accepted provided they are within safe limits. Over-drying damages the physical shape and reduces the nutri-ent content, along with completely spoiling the original color and texture. Physical changes also occur during the process of safe drying in the form of shrinking, puffing, etc., and chemical changes can affect color, taste, odor, etc., (Babu et al. 2018).

When the fresh crop is exposed to heat from any source, the external sur-face of the product becomes hot. This leads to the diffusion of heat into a deeper level of the product, normally by the conduction mode of heat trans-fer. Moisture transfer occurs from the internal surface of the product to the outer surface by various processes, such as moisture diffusion and liquid diffusion, vapor diffusion, etc. (Saeidpour and Wadsö, 2016). Extraction of moisture from the outer surface of the product takes place by the process of evaporation, which takes place through heat transfer. This may be natural convection, forced convection, or both (Bourdoux et al., 2016). The crop can be dried either in thin layer drying (as shown in Figure 2.1(a)) or deep bed drying (as shown in Figure 2.1(b)). The moisture content of the crop decides which method of drying is selected.

To understand the complete process of drying, these important concepts must be considered very carefully: namely, wet moisture content, dry mois-ture content, moisture balance content of moisture content models, falling and constant drying times, thin layer drying rate, deep bed drying, drying rate, dryer efficiency and drying constant (Babalis et al., 2017).

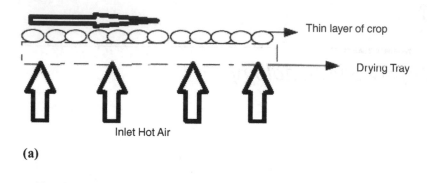

(a)

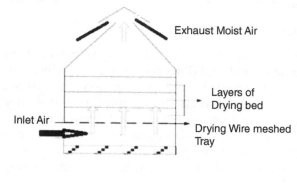

(b)

FIGURE 2.1
Schematic diagram of the (a) thin layer drying, (b) deep bed drying.

2.2 Moisture Content

The drying performance for all crops in any drying condition is evaluated by the wet and dry moisture content. It can be evaluated as follows (Tiwari et al., 2016)

The wet basis moisture content is calculated as

$$MC_{wb} = \frac{M_o - M_d}{M_o} \qquad (2.1)$$

The dry basis moisture content is evaluated as

$$MC_{db} = \frac{M_o - M_d}{M_d} \tag{2.2}$$

Here M_d and M_o are the weight of dry and fresh material, respectively. This differs from crop to crop.

2.3 Moisture Movement Mechanism

Drying can be divided into two types, namely the constant rate period and the falling rate period, depending on the drying rate.

During the constant rate period, the removal of moisture occurs by the process of evaporation on the surface of the crop. Here moisture is evaporated but the drying rate remains constant. The movement of internal moisture is rapid so that the saturated surface is maintained, because the constant drying rate is being maintained. In such a drying methodology, the removal of

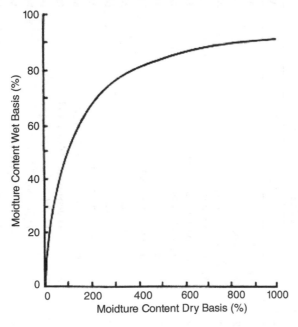

FIGURE 2.2
Graphical relationship between moisture content on a wet and dry basis (Ekechukwu, 1999).

TABLE 2.1

Various Prominent thin Layer Drying Models

S. No.	Name of Model	Mathematical Expression of Model	Root
1	Lewis/Newton	$MR=exp(-kt)$	Newton's law of cooling
2	Page Model	$MR=exp(-kt^n)$	Newton's law of cooling
3	Modified Page Model-I	$MR=exp(-kt)^n$	Newton's law of cooling
4	Modified Page Model-II	$MR=exp-(kt)^n$	Newton's law of cooling
5	Henderson and Pabis model	$MR=aexp(-kt)$	Flick's second law of diffusion
6	Logarithmic model	$MR=aexp(-kt)+c$	Flick's second law of diffusion
7	Midilli Model	$MR=aexp(-kt^n)+bt$	Flick's second law of diffusion
8	Modified Midilli Model	$MR=exp(-kt^n)+bt$	Flick's second law of diffusion
9	Demir et al. model	$MR=aexp(-kt)+b$	Flick's second law of diffusion
10	Two Term Model	$MR=aexp(-k_1t)+bexp(-k_2t)$	Flick's second law of diffusion
11	Two-term exponential model	$MR=aexp(-kt)+(1-a)exp(-akt)$	Flick's second law of diffusion
12	Modified Two-term exponential model	$MR=aexp(-kt)+(1-a)exp(-gt)$	Flick's second law of diffusion
13	Modified Henderson and Pabis model	$MR=aexp(-kt)+bexp\ (-gt)+cexp(-ht)$	Flick's second law of diffusion
14	Thompson Model	$t=aln(MR)+b[ln(MR)]^2$	Empirical Model
15	Wang and Singh model	$MR=1+bxt+axt^2$	Empirical Model
16	Kaleemullah and Kailappan model	$MR=exp(-cxT)+bxt^{(pT-n)}$	Empirical Model
17	Prakash and Kumar model (Prakash and Kumar, 2014)	$MR=at^3+bt^2+ct+d$	Empirical Model

moisture is independent of the movement of the internal moisture. It mainly depends on the removal of large quantities of unbound water from the drying product and is only influenced by it surrounding factors such as temperature, relative humidity, external heat and mass transfer coefficients, and exposed surface area (Srikiatden and Roberts, 2007).

This process continues as long as moisture evaporated from the external surface of the product is equal to moisture supplied to the external surface of the product. After this point, drying rates become progressively slower (Puyate and Lawrence, 2006; El-Sebaii and Shalaby, 2012).

In any drying process, at first the constant drying rate occurs for a time. Then there is mismatch of evaporation of moisture compared to internally supplied moisture. In fact, the amount of supplied moisture is reduced by an amount compared to the amount of evaporated moisture. At that time, the second drying rate begins, called the falling drying rate. As the name implies, the drying rate is reducing over time. The falling drying rate can be further divided into two levels, namely "the first falling drying rate (FFDR)" and "the second falling drying rate (SFDR)". In the process of the falling drying rate, first the FFDR process takes place, and after that SFDR process begins. Table 2.1 presents the some of the prominent thin layer drying models that are frequently used by researchers and scientists.

2.4 Equilibrium Moisture Content (EMC)

Equilibrium moisture content is the least moisture content of the crop. This type of moisture content is also called the safe moisture content of the hygroscopic crop (Kant et al., 2016). During the process of drying, a stage comes when the reduction of the crop weight reaches stagnation. Here, while the crop is in the dryer in favorable circumstances, no moisture removal from the crop takes place. The moisture content in the crop in this situation has reached the equilibrium moisture content (EMC) (Hasburgh et al., 2019). The EMC depends on the amount of previous moisture. The relative humidity of the surrounding of the crop when it attains the safe moisture content is called equivalent relative humidity (ERH). Many theories and reasons for safe moisture content have been documented in the literature by numerous researchers. Figure 2.3 shows the variation EMC with relative humidity.

2.4.1 EMC Models

Various researchers have studied a number of semi-theoretical, theoretical, and empirical models to evaluate the relationship between ERH and EMC

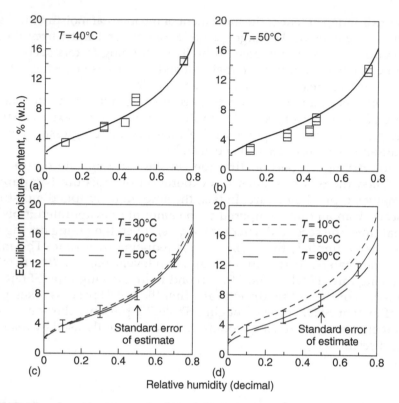

FIGURE 2.3

Equilibrium moisture content percentage versus relative humidity (Zanoelo, 2005).

for isotherms moisture equilibrium of agricultural produce. However, there is no such equation accepted universally that describes moisture balance isotherms for all cultures created (Gocho et al., 2000; Tomar et al., 2017).

Below are some of the ERH/EMC equations.

2.4.1.1 Henderson Equation

The Henderson equation (Chen and Morey, 1989) is the flexible model for explaining the relationship between ERH (ϕ_e) and EMC (M_e) at a specific temperature

$$1 - \phi_e = e^{-kTm_e^n} \tag{2.3}$$

Where n and k depend on temperature T (in K) and differ from crop to crop, ERH (ϕ_e) in decimal, and EMC (M_e) on a dry basis decimal

2.4.1.2 Chung–Pfost Equation

The Chung–Pfost equation established the relationship between EMC and ERH at a certain temperature:

$$\phi_e = e^{-\frac{a}{T+b}\left(e^{-cM_e}\right)} \tag{2.4}$$

Where a, b and c are constants. These constants are dependent on individual crops. Hence they vary on a crop to crop basis.

2.4.1.3 Modified Halsey Equation

The Modified Halsey equation developed by Iglesias and Chirife (1976) also proposed a relationship between EMC and ERH which is as follow:

$$\phi_e = e^{-e^{-(a+bT)}M_e^{-c}} \tag{2.5}$$

Where a, b, c are constants also dependent on the crop.

2.4.1.4 Modified Oswin Equation

The Modified Oswin equation also developed a relationship between EMC and ERH, as follows:

$$\phi_e = \left[\left(\frac{a+bT^d}{M_e}\right)^c + 1\right]^{-1} \tag{2.6}$$

Where a, b, and c in Equations 2.4, 2.5 and 2.6 are constants varying from crop to crop.

2.5 Drying Theory

There are two type of product used in drying, namely (a) hygroscopic/agricultural products; and (b) non-hygroscopic products (textiles, sand, soil, paper, or dust). Hygroscopic materials are a mixture of bound and unbound water content/moisture, while non-hygroscopic materials have only unbound water content/moisture. Equilibrium vapor pressure at a specific temperature is exerted by bound moisture, which is lower than pure water in the case of unbound moisture; the equilibrium vapor pressure is the same as pure water (Abou et al., 2019).

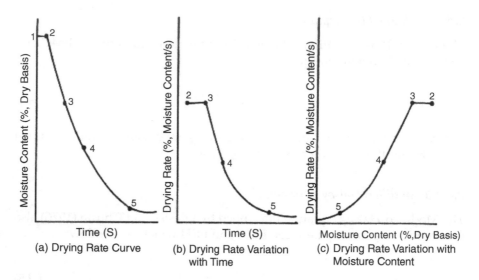

FIGURE 2.4

Schematic presentation of the drying curve where 1–2 is heating time, 2–3 is constant drying rate time/period, 3 is the point of critical moisture content, 3–4 the first falling rate and 4–5 the second falling rate (Ekechukwu, 1999).

During the process of drying, the product to be dried is heated by renewable or non-renewable energy sources. With the application of heat, the vapor pressure of the product increases at constant moisture content. The moisture transfer from the product to its surroundings takes place when the surrounding vapor pressure is lower than the product's vapor pressure. The rate of moisture transfer from the product is directly proportional to the vapor pressure gradient between product and surroundings. The drying rate can be defined as the removal of moisture in per unit time. See Figure 2.4 for a schematic diagram of the drying curve.

In the curve, there are four different processes, namely the heating period, the constant drying rate period, the first falling rate and the second falling rate. All the crops go through these stages during the process of drying. Figure 2.4 shows the variation of moisture content and drying rate in food drying curves.

2.6 Drying Rate Equation

During the process of drying, the drying rate is classified into two types, namely the constant drying rate period, and the falling drying rate period (Defraeye and Radu, 2017).

2.6.1 Constant Drying Rate Period

In this drying rate period, moisture evaporation occurs on the surface of a material. This process is like the evaporation of water from the free-water surface. The process depends on the surrounding conditions (Kant et al., 2016). After the constant drying rate is maintained for a certain span of time, the falling drying rate begins until the crop attains equilibrium moisture content. This procedure of drying is practical for hygroscopic materials. However, for non-hygroscopic products, drying means the constant drying rate only and no falling drying rate.

The drying rate depends on a range of conditions such as the difference in vapor pressure between the wet surface and the drying surface, the surface area, the drying air velocity, and the mass transfer coefficient. Figure 2.5 shows the variation of the drying rate with free moisture. There are three major sections, namely the heating zone, the constant drying rate and the falling drying rate. In the graph, points A to B represent the heating zone of the material, points B to C represents the constant drying rate, C to D is the first falling rate, and D to E the second falling rate.

2.6.2 Falling Drying Rate Period

The falling drying rate is the next step after constant drying rate. The process begins when the product moisture content becomes less than the critical moisture content.

This process is also further divided into two sections, namely the first drying rate period, and the second drying rate period. The first drying rate

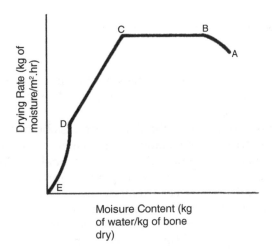

FIGURE 2.5
Variation of drying rate along with free moisture.

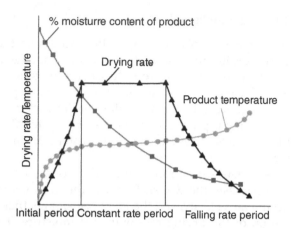

FIGURE 2.6
A schematic presentation of various food graphs (Lamidi et al., 2019).

period relates to unsaturated surface drying, and the second drying rate period takes place with inner moisture transfer to the outer surface. This happens through the process of diffusion and whole process is a slow one. This is the most critical factor in the drying process (Gunathilake et al., 2018). However, initial moisture content is also an important factor in the drying process. Figure 2.6 presents various food graphs showing percentage moisture content, drying rate, and product temperature.

2.7 Shrinkage

In the drying process, significant changes happen in the external features of the material, such as in volume, area, length, thickness, etc. This phenomenon is called shrinkage.

Shrinkage plays a vital role in the drying process because it is an important quality of the dried product, along with changes in form, an increase in hardness, and a change in volume (El-Sebaii and Shalaby, 2012; Mayor and Sereno, 2004). In most cases, it creates a negative impression with consumers (Mahiuddin et al., 2018). However, consumers expect some traditional dried products to be shrunken and wrinkled—for example, dried apples, raisins, etc. If the shrinkage is not uniform, unbalanced stress is created, and then surface cracking takes place. Shrinkage also happens as a result of the reduction of the moisture content in the dried product.

Below are the important parameters related to shrinkage.

2.7.1 Shrinkage Model

This describes a reduced dimensional increase in thickness area and volume. D_s is the dimension of shrinkage (volume, area, thickness).

There are two models for evaluating this parameter.

Model 1: Linear Empirical Model:

$$D_s = a_1 M + a_2 \tag{2.7}$$

Where a_1 and a_2 are constant and M is moisture content.

Model 2: Non-linear Empirical Model:

$$D_s = 0.16 + 0.816 \frac{M}{M_o} + 0.022 \exp\left(\frac{0.018}{M + 0.025}\right) + 0.209 - \left(\frac{0.966}{M_o + 0.796}\right)\left(1 - \frac{M}{M_o}\right) \tag{2.8}$$

2.7.2 Volume Shrinkage

This is also known as relative volumetric shrinkage. This is the ratio between final volume per initial volume of the product. It is denoted by S_v, which is equal to V/V_0.

$$S_v = \frac{V}{V_0} = \left(\frac{M + 0.8}{M_o + 0.8}\right); \quad S_v = \frac{V}{V_0} = \left(\frac{A}{A_0}\right)^{\frac{3}{2}} \tag{2.9}$$

2.7.3 Bulk Density

The bulk density is defined as the ratio between the current weight of the crop per initial volume of the crop. This is denoted by ρ_{bc}.

$$\rho_{bc} = \frac{m_s + m_w}{V_s + V_w + V_a} \tag{2.10}$$

Where m_w and m_s are the masses of water and dry solids, respectively, and V_a, V_w, and V_s are the volumes of air pores, water, and dry solids, respectively.

2.7.4 Particle Density

The particle density is the ratio between current total mass of the product per overall volume of the product. This is denoted by ρ_{pc}.

$$\rho_{pc} = \frac{m_s + m_w}{V_s + V_w} \tag{2.11}$$

2.7.5 Dry Solids Density

The dry solids density is the ratio of the mass of the dry solid per volume required by that solid dried crop. This is denoted by ρ_{ds}. Mathematically, it is expressed as

$$\rho_{ds} = \frac{m_s}{V_s} \tag{2.12}$$

2.7.6 Equilibrium Density

The equilibrium density is the ratio of mass per volume when crop attains the equilibrium condition with its surroundings during the process of drying. This is represented as

$$\rho_e = \frac{m_e}{V_e} \tag{2.13}$$

Where $V_e = (V_s + V_w + V_a)$

2.7.7 Porosity

The porosity is the ratio between volumes of the air presence in the product per total volume of the product. It is denoted by ε_p. Mathematically, it is expressed as

$$E = \frac{V_a}{V_s + V_w + V_a} \tag{2.14}$$

2.8 Pretreatments for Drying

Fruit and vegetables help those who consume them to gain weight, and they supply minerals, vitamins, and energy. Hence it is very important to preserve the most important fruits and vegetables so that they can be eaten throughout the year, by this means avoiding malnutrition. But to make the drying process effective and retain the foods' nutritive value, proper pretreatment is very important.

2.8.1 Pretreatment of Vegetables

First, for both vegetables and fruit, any foreign material is to be removed from the surface of the crops. After this, a blanching process is applied to all

TABLE 2.2

Pretreatment for Drying of Selected Vegetables

Vegetables	Preparations (After Washing and Cutting to Suitable Size)	Blanching Time (min) with Steam	Cooling Time (min) With Cool Water
Asparagus	As normal for cooking	4–6	4–5
Green beans	Cut in pieces/strips	2–3	2
Beets	Peel	Included in cooking	Included in cooking
Broccoli	Trim, and quarter stalks length-wise	3–4	2
Brussels sprouts	Cut in half lengthwise	7–8	5–6
Cabbage	Remove outer leaves, cut in strips 0.32 cm thick.	3	2
Carrots	Cut roots in strips of 0.32 cm thickness	3–4	4
Cauliflower	Trim stalks and cut into florets	5–6	4–5
Celery	Slice stalks	2–3	2–3
Corn	Husk, trim, blanch until 'milk' in corn is set	3–5	3
Mushrooms	Slice tender stalks 0.64 cm thick	None	None
Onions	Remove tops and root ends, cut into slices 0.32–0.64 cm thick	None	None
Parsley and other herbs	Wash and remove foreign materials	None	None
Peas	Wash and remove foreign materials	3–4	3
Peppers	Cut into strips or rings of 0.64–1.28 cm thickness	None	None
Potatoes	Peel and cut into strips of 0.64 cm thickness, or 1.28 cm thick slices	7–9	6–7
Spinach and other greens	No special treatment	2–3 (until wilted)	2
Squash	1. Remove seeds and cavity pulp 2. Cut into 2.54 cm wide strips, 1.28 cm thick	3	1–2
Squash	Cut into 0.64 cm thick slices	3	1–2
Tomatoes	Slice 1.28 cm thick	None	None

vegetables except onions, peppers or spices. This pretreatment process saves time and preserves color and vitamins, mainly through the relaxation of the inner tissues of the crop. A detailed analysis of the pretreatment of a range of vegetables is presented in Table 2.2.

2.8.2 Pretreatment of Fruit

Some fruits do not need pretreatment before drying, but some others do. Cherries, seedless grapes, melons, prunes and plums do not need pretreatment, but it is needed for apples, apricots, peaches, nectarines and pears. Table 2.3 presents some selected fruits with their pretreatment procedures.

TABLE 2.3

Pretreatment for the Drying of Selected Fruits

Fruit	Preparation (Washing)	Pre-treatment
Apple	1. Cut into rings or slices 0.32–0.64 cm thick 2. Coat with ascorbic acid solution (uses 2¼ tsp/cup water)	Either: Soak for 5 min in sodium sulfite solution Or: Steam-blanch for 3–5 min, depending on size and texture
Apricots (firm, fully ripe)	1. Cut in half and remove pit (do not peel) 2. Coat with ascorbic acid solution (1 tsp/cup)	Either: Soak 5 min in sodium sulfite solution Or: Steam blanch 3–5 min
Bananas (firm, ripe)	Peel and cut into 0.32 cm thick slices	No treatment necessary (may be dipped in lemon juice)
Berries (firm)	Cut in half	No special requirements
Cherries (fully ripe)	Wash Remove stems and pits	No special requirements
Grapes (seedless varieties)	1. Sort, leave whole on stems in small bunches	No special requirements
Melons (mature, firm and heavy)	1. Remove outer skin, any fibrous tissue and seeds 2. Slice 0.64–1.28 cm thick	No pretreatment necessary
Nectarines and Peaches (ripe, firm)	1. Peel, cut in half and remove pit/quarters/slices 2. Coat with ascorbic acid solution (1-tsp/cup)	Either: Soak 5–15 min in sodium sulfite solution Or: Steam blanch halves 8–10 min, slices 2–3 min
Pears	1. Cut in half lengthwise/quarters or eighths, or slice 0.32–0.64 cm thick 2. Coat with ascorbic acid solution (1 tsp/cup)	Either: Soak 5–15 min sodium solution Or: Steam blanch 5–7 min
Plums and prunes	Cut large fruit into halves (pit removed) or slices	Steam blanch halves or slices 5–7 min

Problems

2.1 Explain the drying rate and its classification.

2.2 Discuss the various models of equilibrium moisture content.

2.3 Discuss the importance of the pretreatment of fruit and vegetables.

2.4 Discuss the various type of moisture content and their relationships.

2.5 Discuss shrinkage and its empirical relations in different conditions.

References

Abou, M.M.N., Boukar, M. and Madougou, S., 2019. Effect of drying air velocity on drying kinetics of tomato slices in a forced-convective solar tunnel dryer. *Journal of Sustainable Bioenergy Systems*, 9(2), 720–726.

Babalis, S., Papanicolaou, E. and Belessiotis, V., 2017. Fundamental mathematical relations of solar drying systems. In Prakash, O. and Kumar, A. (eds) *Solar Drying Technology* (pp. 89–175). Singapore: Springer.

Babu, A.K., Kumaresan, G., Raj, V.A.A. and Velraj, R., 2018. Review of leaf drying: Mechanism and influencing parameters, drying methods, nutrient preservation, and mathematical models. *Renewable and Sustainable Energy Reviews*, 90, 536–556.

Bala, B.K. and Bala, B.K., 1997. *Drying and Storage of Cereal Grains*. Enfield (NH): Science Publishers.

Bourdoux, S., Li, D., Rajkovic, A., Devlieghere, F. and Uyttendaele, M., 2016. Performance of drying technologies to ensure microbial safety of dried fruits and vegetables. *Comprehensive Reviews in Food Science and Food Safety*, 15(6), 1056–1066.

Chauhan, P.S. and Kumar, A., 2016. Performance analysis of greenhouse dryer by using insulated north-wall under natural convection mode. *Energy Reports*, 2, 107–116.

Chen, C.C. and Morey, R.V., 1989. Comparison of four EMC/ERH equations. *Transactions of the ASAE*, 32(3), 983–0990.

Defraeye, T. and Radu, A., 2017. Convective drying of fruit: A deeper look at the air-material interface by conjugate modeling. *International Journal of Heat and Mass Transfer*, 108, 1610–1622.

Ekechukwu, O.V., 1999. Review of Solar energy drying system I: An overview of drying principles and theory. *Energy Conversion and Management*, 40, 593–613.

El-Sebaii, A.A. and Shalaby, S.M., 2012. Solar drying of agricultural products: A review. *Renewable and Sustainable Energy Reviews*, 16(1), 37–43.

Gocho, H., Shimizu, H., Tanioka, A., Chou, T.J. and Nakajima, T., 2000. Effect of polymer chain end on sorption isotherm of water by chitosan. *Carbohydrate Polymers*, 41(1), 87–90.

Gunathilake, D.M.C.C., Senanayaka, D.P., Adiletta, G. and Senadeera, W., 2018. Drying of agricultural crops. In Chen, G. (ed) *Advances in Agricultural Machinery and Technologies* (pp. 331–365). Boca Raton, FL, USA: CRC Press.

Hasburgh, L.E., Craft, S.T., Van Zeeland, I. and Zelinka, S.L., 2019. Relative humidity versus moisture content relationship for several commercial wood species and its potential effect on flame spread. *Fire and Materials*, 43(4), 365–372.

Iglesias, H.A. and Chirife, J. (1976) Prediction of the effect of temperature on water sorption isotherm of food materials. *Journal of Food Technology*, 11(2, March–April), 109–116.

Kant, K., Shukla, A., Sharma, A., Kumar, A. and Jain, A., 2016. Thermal energy storage based solar drying systems: A review. *Innovative Food Science & Emerging Technologies*, 34, 86–99.

Lamidi, R.O., Jiang, L., Pathare, P.B., Wang, Y.D. and Roskilly, A.P., 2019. Recent advances in sustainable drying of agricultural produce: A review, *Applied Energy*, 233–234, 367–385.

Mahiuddin, M., Khan, M.I.H., Kumar, C., Rahman, M.M. and Karim, M.A., 2018. Shrinkage of food materials during drying: Current status and challenges. *Comprehensive Reviews in Food Science and Food Safety*, 17(5), 1113–1126.

Mayor, L. and Sereno, A.M., 2004. Modelling shrinkage during convective drying of food materials: A review. *Journal of Food Engineering*, 61(3), 373–386.

Prakash, O. and Kumar, A., 2014. Environomical analysis and mathematical modeling for tomato flakes drying in modified greenhouse dryer under active mode. *International Journal of Food Engineering*, 10(4), 669–681

Puyate, Y.T. and Lawrence, C.J., 2006. Sherwood's models for the falling-rate period: A missing link at moderate drying intensity. *Chemical Engineering Science*, 61(21), 7177–7183.

Saeidpour, M. and Wadsö, L., 2016. Moisture diffusion coefficients of mortars in absorption and desorption. *Cement and Concrete Research*, 83, 179–187.

Srikiatden, J. and Roberts, J.S., 2007. Moisture transfer in solid food materials: A review of mechanisms, models, and measurements. *International Journal of Food Properties*, 10(4), 739–777.

Tiwari, S., Tiwari, G.N. and Al-Helal, I.M., 2016. Development and recent trends in greenhouse dryer: a review. *Renewable and Sustainable Energy Reviews*, 65, 1048–1064.

Tomar, V., Tiwari, G. and Norton, B., 2017. Solar dryers for tropical food preservation: Thermophysics of crops, systems and components. *Solar Energy*, 154, 2–13.

Zanoelo, E.F., 2005. Equilibrium moisture isotherms for mate leaves. *Biosystems Engineering*, 92(4), 445–452.

3

Various Designs of Solar Drying Systems

3.1 Introduction

Solar energy is the source of all form of energy apart from nuclear. However, despite it being abundantly available, it is present in a dilute form. Hence some conversion devices are required to employ this solar energy in a meaningful manner apart from its uses for lighting and heating. Thermal energy can be obtained from solar radiation by using solar collectors, and direct current electricity can be converted from solar radiation via photovoltaic panels. Sometimes both functions are carried out simultaneously, but with lesser efficiency. One prominent application of solar thermal energy is the drying of agricultural produce (Kannan and Vakeesan, 2016).

This section of the book includes a detailed discussion of the solar drying system. Based on the mechanism of drying methodology, it is classified broadly into three types: namely, open-air solar drying, direct solar drying, and indirect solar drying. Each type will be discussed in detail in the following sections.

3.1.1 Open Sun Drying (OSD)

Open sun drying is also called natural drying. This method of drying has been used throughout the world for generations. This type of drying is shown in Figure 3.1.

In open sun drying, the short wavelengths of solar radiation fall on the surface of the crop, which lies on the Earth's surface under the open sky. Some part of the solar radiation is reflected on to the surrounding, and the remaining part is absorbed by the crop. The amount of solar radiation absorbed by the crop depends on the color and porosity of the crop/agricultural produce. The reflected solar radiation does not play any part in the drying, but the absorbed radiation plays a critical role in the process. Because of this absorbed solar radiation, the temperature of the agricultural produce starts to increase, mainly as a result of the transformation of solar energy to thermal

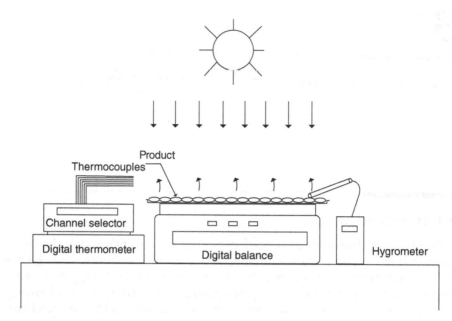

FIGURE 3.1
Schematic diagram of open sun drying (OSD) (Akpinar, 2006).

energy. Hence it is crucial to increase the amount of solar radiation absorbed by the agricultural produce that is undergoing open sun drying (Sahu et al., 2016). In this type of drying, the convection mode of heat transfer is prominent, compared to conduction and radiation modes. The convective heat transfer happens between the crop surface and the ambient air. Mass transfer also happens between them through moist air. Hence the crop becomes dry after the removal of excess moisture. The removal of moisture takes place by the process of evaporation.

In the initial phase of drying, moisture removal is rapid, and as time passes, it continues to decrease at a fast rate. It is somewhat like an exponential decrease for crops with a high moisture content crop. For the other two types of crop, namely low- and medium-moisture content crops, it is not an exponential decrease. Ultimately, a minimal loss of moisture happens, and or practical purposes, there is no moisture loss. Then the drying process is accepted as complete.

There is a significant disadvantage to open sun drying, mainly from the serious effects of the presence of rodents, wind, dirt, birds, insects, and microorganisms. Sudden rain or a storm worsens the situation even further. Moreover, over/under drying, contamination with foreign materials, and ultraviolet (UV) radiation discoloration are problems in open sun drying (Pandey et al., 2018). Over time, because of changes to the climate, such problems are becoming more problematic. Hence the open sun drying method is not the preferred method because it does not meet the standard specifications and quality control requirements of nationally and internationally approved standards.

3.1.2 Direct Solar Drying (DSD)

Direct solar drying (DSD) is a development of open sun drying. Here all the problems related to natural or open sun drying are rectified, such as those related to weather and dirt. The dryable product is kept in a protected area away from contamination. The product is kept in a drying chamber, protected on four sides by the walls. As the energy is being received from the sun, a transparent cover is used. Cabinet dryers and greenhouse dryers (shown in Figures 3.2 and 3.3) are the most common examples of direct solar dryers.

In this method of drying, the actual governing effect is the greenhouse effect. By this alone drying takes place. In the drying chamber, shortwave solar radiation enters through the transparent wall and is absorbed by the inner surface and the crop (Agrawal and Sarviya, 2016). After the absorption of the solar radiation, the surface begins to emit long wavelength radiation, but this is restricted by the transparent wall. This causes the temperature inside the chamber to begin to rise, and the relative humidity decreases gradually. This is the most favorable condition for drying. Direct solar drying can be operated either by natural convection or forced convection modes of heat transfer (Selvaraj et al., 2018). Dryers driven by the natural convection mode of heat transfer have a low drying rate; but the forced convection mode of heat transfer has a high drying rate.

3.1.3 Indirect Solar Drying (ISD)

The indirect solar dryer is the next step in direct solar drying. A photograph of an indirect solar dryer is presented in Figure 3.4. Here the crop is not directly exposed to the solar radiation. The main reason for this is that crop quality and color deteriorate with direct exposure to the solar radiation (Slimani et al., 2016). The dryable crop is placed in the drying chamber, a closed box with inlet and outlet openings. Hot air from a solar air heater enters through the inlet opening. The air heats the crop and passes out through the other opening as humid air. This takes moisture away from the crop. This method of drying is used in thin layer drying methodology. The indirect solar dryer can also operate in both modes of heat transfer, namely the natural convection mode and the forced convection mode. However, in practice, natural convection based indirect solar dryers are not used, because of the very low flow rate of the incoming air. Hence indirect solar dryers under the forced convection mode are being used for all practical purposes. In these dryers there are two main components, namely the solar air heater and the drying

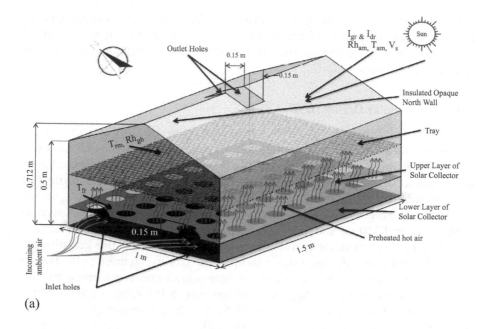

(a)

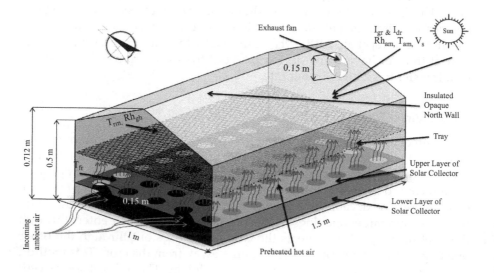

(b)

FIGURE 3.2
Schematic diagram of solar greenhouse drying (DSD) (Chauhan and Kumar, 2018).

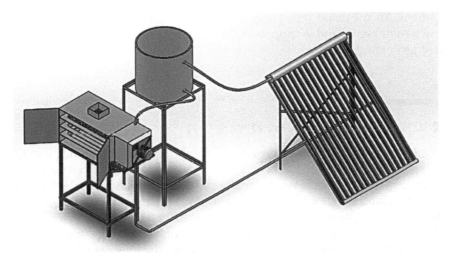

FIGURE 3.3
Schematic diagram of cabinet solar dryer (Iranmanesh et al., 2020).

FIGURE 3.4
Photograph of an indirect solar dryer (ISD) (Essalhi et al., 2018).

chamber. An insulated pipe connects the two, through which hot air comes from a solar heater and is supplied to the drying chamber, which is well insulated against thermal loss.

3.2 Classification of Solar Dryers

Solar energy is being used extensively in food preservation by the process of drying. Since this method has been used for generations, there has been much improvement in it over time. Therefore the whole process of solar drying is being classified as follow (as shown in Figure 3.5). In a broad sense, it is classified into two groups, namely open sun drying and controlled solar drying.

Open sun or natural drying depends mainly on the average intensity of the solar radiation, the average ambient temperature, the average relative humidity, and the moisture content of the agricultural produce (Huddar and Kamoji, 2019). A more practical way to use solar radiation from the sun for crop drying has been developed, called controlled solar drying (Chauhan et al., 2017). This is a modified version of open sun drying, with its problems or disadvantages rectified in this dryer. Controlled dryers or modern-day solar food dryers are advanced compared to traditional open-air sun dryers in five significant ways: (i) quicker; (ii) more effective; (iii) hygienic; (iv) healthier; and (v) cheaper.

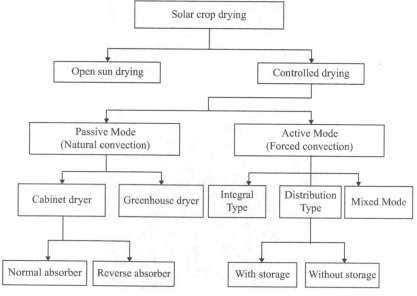

FIGURE 3.5
Classification of solar drying methodology (Sharma et al., 2009).

Modern solar drying systems have modified the shape, size and ways of working so that maximum advantage can be extracted from the available resources. Since drying is a complicated heat and mass transfer process, it can take place in both operating modes, namely the natural convection mode and the forced convection mode of heat transfer. Open sun drying operates only in the natural convection mode of heat transfer, and in this process the drying rate is very much less. However, in order to increase the drying rate, a controlled dryer or modern dryer can be operated in both modes of heat transfer (Yadav et al., 2018). Dryers based on the natural convection mode of heat transfer are known as passive solar dryers, and forced convection mode based solar dryers are known as active solar dryers. Solar air heaters play an essential role in assisting the indirect and mixed modes of the solar dryer. Active solar dryers are equipped with air blowers that run on electricity powered by solar energy to produce forced convection mode heat transfer.

3.3 Passive Solar Dryers

Passive solar dryers are often referred to as natural ventilation solar dryers, where no external source of energy is required to run the dryer. In this dryer, incoming air is heated up with the help of solar energy. The hot air then passes over the crop's surface. The thermosyphon effect combined with the greenhouse effect are the main operators for natural ventilation. This makes these solar dryers an efficient and desirable choice for farmers in remote areas (Jain and Tiwari, 2004). Active mode (integral form), indirect mode (distributed form), and mixed-mode are the three basic types of passive solar dryers.

3.3.1 Direct Passive Solar Dryers

In a direct passive solar dryer, the crop is put in a transparent-walled drying chamber. Solar radiation beams directly on to the crop through the transparent cover, is converted into thermal energy by the greenhouse effect, and increases the temperature of the product. This leads to the removal of moisture from the crop and reduces the inside relative humidity. The moisture is evaporated by the moisture diffusion process.

3.3.2 Cabinet Dryers

Cabinet dryers have been used for drying crops for a long time. As early as 1957, research was being carried out on certain types of dryers. The Brace Research Institute, McGill University, Canada, reported groundbreaking

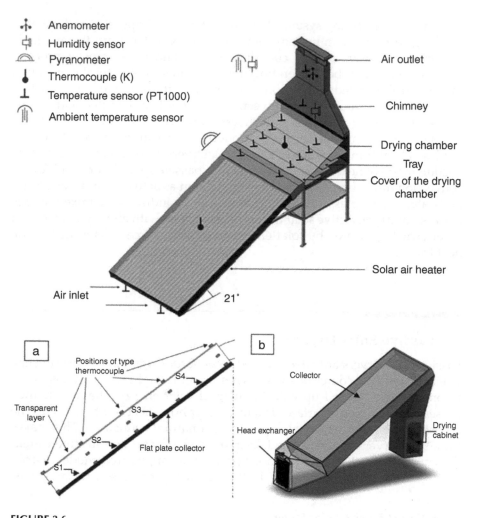

FIGURE 3.6
(a) Solar cabinet dryer in natural convection mode (Abubakar et al., 2018). (b) Cabinet solar dryer in forced convection mode (Motahayyer et al., 2019).

research on solar cabinet dryers. Numerous other passive cabinet solar dryers were built and tested for a diversity of crops in different locations.

Cabinet solar dryers are small in capacity, which suits domestic purposes. In Figures 3.6(a) and 3.6(b), such dryer operated in natural convection and forced convection mode of heat transfer are shown.

The top cover of the cabinet solar dryer is either single- or double-glazed to receive energy from solar radiation. Figure 3.7 shows the mixed-mode-based solar cabinet dryer assisted by a chimney. The chimney

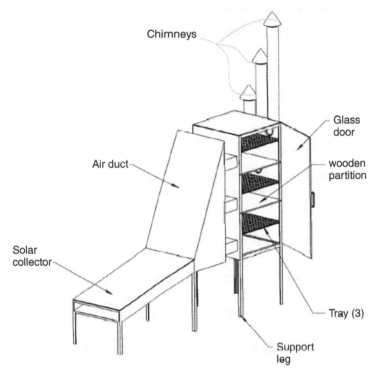

FIGURE 3.7
A natural circulation solar cabinet dryer under mixed-mode operation (César et al., 2020).

encourages a uniform rate of moist air leaving the drying chamber by creating a suitable draft.

The basic guidelines recommended by the Brace Research Institute for the construction of solar cabinet dryers are as follows:

(i) Cabinet length should be at least three times dryer size.

(ii) An optimal angle of slope as a function of the local latitude for the glazing.

(iii) The inside walls are to be painted black.

(iv) To ensure a proper air circulation inside the drying chamber, the drying trays are positioned above the cabinet floor.

(v) Double glazing top cover protects against UV radiation and prevents deterioration.

(vi) The dryer should be made from good quality locally available material.

A cabinet dryer is very efficient compared with open sun dryers, reducing the drying time by half. Sodha et al. (1985) compared the performance of cabinet solar dryers with open sun drying, and their results revealed that the drying efficiency of the cabinet solar dryer is greater and offers better product quality.

Solar cabinet dryers can be classified based on the placement of the absorber plate and incident solar radiation, as below.

3.3.2.1 Normal Absorber Cabinet Dryer

The normal absorber cabinet dryer is the basic structural form of the cabinet dryer. Though it is much better than the open sun dryer, it has the following significant disadvantages:

(i)   This type of dryer is used for small batches, and its use is limited to small-scale applications.

(ii) Crop discoloration occurs because of the direct solar radiation exposure.

(iii) Transmissivity of the cover glass is reduced over time because of condensation.

(iv) The uneven rise in the crop temperature affects the moisture evaporation rate.

(v)   Limited use of variable absorber plate coatings.

3.3.2.2 Reverse Absorber Cabinet Dryer (RACD)

The reverse absorber cabinet dryer (RACD) is an improved version of the normal absorber cabinet dryer. With this dryer, the solar radiation is allowed to fall on the absorber plate from below the surface, to reduce the convective heat loss. This prevents discoloration of the crop because it is not exposed directly to solar radiation. The crop is placed on a wire mesh tray. The absorber plate, facing downward, is placed at a sufficient distance (0.05 m) from the lower surface. A cylindrical reflector is placed under the absorber surface, its aperture fitted with a glass cover. This reduces convective heat loss from the absorber plate. The glass cover is inclined at a 45° angle to obtain full solar radiation incidence on the absorber plate, as shown in Figure 3.8. The area of the glass cover is equal to the lower area of the drying chamber. The cylindrical reflector reflects the solar radiation toward the absorber plate. A portion of solar radiation is lost through the glass cover and the remainder is passed into the air through convection heat transfer. The circulation of heated air passes through the crop placed in the drying chamber, removing its moisture content with the hot air. This hot air becomes humid because of the moisture evaporation and is released through the vent.

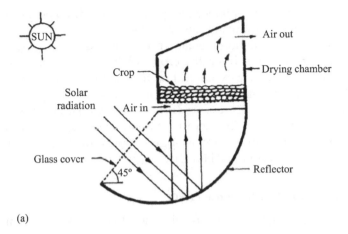

(a)

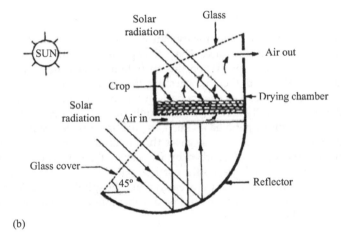

(b)

FIGURE 3.8
Schematic diagram of a reverse absorber cabinet dryer (a) with glass, and (b) without glass (Sharma et al., 2009).

3.3.3 Greenhouse Dryers

Open sun drying is the traditional method of preserving agricultural produce in developing countries (Essalhi et al., 2018). This mode of drying
under intimidating weather conditions contributes to a significant degradation in the quality and quantity of the product (Huddar and Kamoji, 2019).
These losses are mainly a result of contamination by dirt and dust, and damage by insects, rodents and domestic animals, so it is essential to seek alternative means of drying the agricultural produce cheaply. Solar greenhouse
dryers emerge as the solution to the shortcomings of open sun drying. They

decrease product loss and enhance the quality of the dried crops in comparison to open sun drying (Jain and Tiwari, 2004).

The structure of a greenhouse solar dryer is the same as a regular greenhouse. In the greenhouse dryer, the crop is laid in trays, and the removal of moisture happens either by natural or forced convection; which of these applies depends on the mode of removal of exhaust air. The potential advantages of greenhouse solar dryers are (i) they are more efficient in the use of the received solar energy; (ii) there is comparatively low initial inversion; and (iii) the structure can be used as a greenhouse or dryer throughout the year, which provides economic benefits.

3.3.3.1 Definition of Greenhouse Effect

Encyclopedia 2000 defines the greenhouse effect as follow (Kannan and Vakeesan, 2016)

> Greenhouse effect, a term for the role the atmosphere plays in insulating and warming the earth's surface. The atmosphere is largely transparent to incoming solar radiation. When this radiation strikes the earth's surface, some of it is absorbed, thereby warming the earth's surface. The surface of the earth emits some of its energy back out in the form of infrared radiation. As this infrared radiation travels through the atmosphere, much of it is absorbed by atmospheric gases such as carbon dioxide, methane, nitrous oxide, and water vapour. These gases then re-emit infrared radiation, some of which strikes and is absorbed by the earth. The absorption of infrared energy by the atmosphere and the earth is called the greenhouse effect, and maintains a temperature range on earth that is hospitable to life. Without the greenhouse effect, the earth would be a frozen planet with an average temperature of about $-18°C$ ($0°F$).

The same greenhouse effect can be accomplished on a micro-level when a building has transparent walls and roof. Glass and Perspex are main transparent material used for such walls and roofs. When the short wavelengths of solar radiation pass through the walls and roof, they are absorbed by the objects inside, which in turn are warmed. The warmed objects emit long-wavelength radiation, which cannot get out escape because of the transparent walls and roof, and this causes an increase in temperature in the building. Such a building is known as a greenhouse. It maintains a required controlled environment, so it is also called a controlled environment greenhouse.

The limitations of cabinet dryers are capability, condensation of moisture, amount of crop, and cost. These are addressed in greenhouse dryers, which are also a type of direct solar dryer. They are mainly used for bulk-level drying of the crop, as shown in Figure 3.10. Greenhouse dryers are also called tent dryers, upgraded versions of classic greenhouse dryers. In such dryers, the uniform air circulation is very carefully monitored. The hot air is supplied via a pipeline that is fitted and positioned very carefully to regulate air movement. The pipe is also used to cover the transparent wall to provide efficient glazing. The heat storage facility and insulation on top of the glazing at night increase overall performance and minimizes heat losses. The operating principle of the greenhouse dryer and cabinet dryer are identical. However, perfectly built greenhouse dryers provide better control throughout the drying cycle and are suitable for larger-scale drying compared to cabinet dryers.

The broad classification of the greenhouse system is presented in Figure 3.9. Greenhouses are available in various shapes and sizes, based on requirements (Lakshmi et al., 2018; Lingayat et al., 2020).

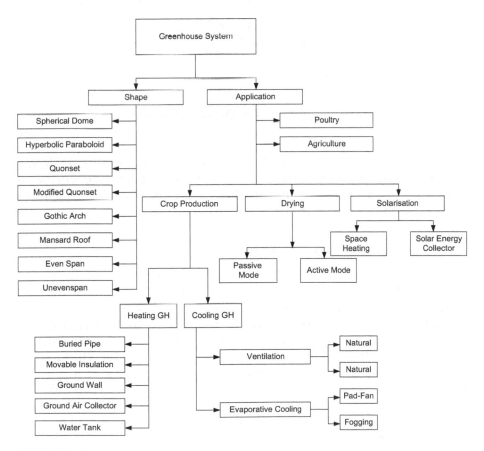

FIGURE 3.9
Classification of greenhouse dryers.

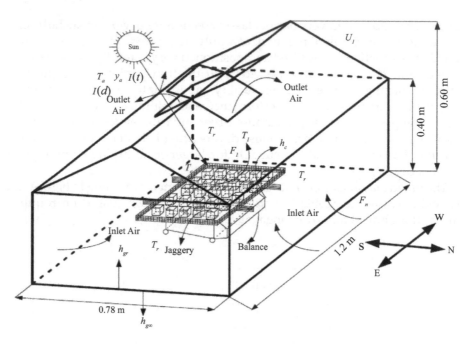

FIGURE 3.10
Schematic diagram of a typical greenhouse dryer (Kumar and Tiwari, 2006).

To carry out thin-layer drying, the crop is spread on the wire mesh tray inside the dryer, as shown in Figure 3.10. Among the different types of greenhouse dryer, even-span roof types provide a better uniform mixing of airflow and air distribution inside the drying chamber.

3.3.4 Indirect Passive Solar Dryers

The indirect passive solar dryer is also a type of indirect solar dryer that works on the principle of the natural convection mode of heat transfer. A diagram of this setup is shown in Figure 3.11. In this system, the incoming air flow rate is very low, hence the rate of heat transfer is also very low. With this dryer, there is no provision of any equipment to enhance the natural heat transfer. Things work naturally in this dryer. The arrangement is the same as for an indirect solar dryer consisting of a solar air heater and drying chamber. In the drying chamber, the crop is dried with the help of hot air, which is provided by the solar air heater and passes out via an overhead vent. In the drying chamber, crops are spread on trays without overlapping.

This dryer is easy is to fabricate, and has a very low operating cost but a major disadvantage is the low drying rate.

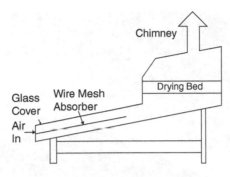

FIGURE 3.11
Schematic diagram of an indirect passive solar dryer (Madhlopa et al., 2002).

3.3.5 Mixed Mode Passive Solar Dryers

Mixed-mode solar dryers have the advantages of both direct and indirect types of solar dryer. They can be operated in both the natural (passive) and force convection (active) modes of heat transfer. The passive mode of mixed-mode solar dryer airflow depends entirely on the thermosyphon effect combined with the greenhouse effect. Figures 3.12 and 3.13 show the mixed-mode passive solar dryer. Here the airflow rate is quite low and hence the drying rate is also low compared to the mixed-mode active solar dryer. In the mixed

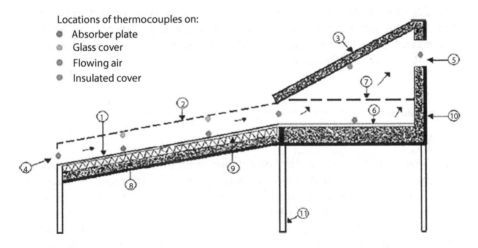

FIGURE 3.12
Schematic diagram of the mixed-mode passive solar dryer: (1) collector absorber plate, (2) collector glass cover, (3) insulated cover, (4) inlet vent, (5) outlet vent, (6) drying chamber absorber plate, (7) wire mesh, (8) insulation, (9) electric heating plate, (10) wooden case, and (11) angle iron stand (Singh and Kumar, 2012).

FIGURE 3.13
Photograph of the mixed-mode passive solar dryer (Singh and Kumar, 2012).

solar dryer, solar thermal energy is received both directly and indirectly—directly in the drying chamber and indirectly through the solar air heater or solar collector. Such dryers can be used for a variety of crops that are suitable for low-temperature thermal drying.

3.4 Active Solar Dryers

After open sun drying, passive solar dryer came into use and then an active solar dryer was developed, working on the principle of the forced convection mode of heat transfer. Fans or a blower are used to increase the airflow rate so that enhanced heat transfer takes place. Such dryers are energy intensive, but the drying rate is significantly higher. These dryers are highly suitable for crops with a high moisture content, and are particularly suitable for low-temperature thermal drying. Sometimes external heaters are also used to preheat incoming air to enhance the heat transfer.

There are three main types of active solar dryers: direct active solar dryers, indirect active solar dryers, and mixed-mode solar dryers.

3.4.1 Direct Active Solar Dryers

Direct active solar dryers have the same structure as a direct passive solar dryers except for the use of a fan/blower. A photograph of a direct active solar dryer is presented in Figure 3.14. A fan or blower can be used either at the inlet or the exhaust point. However, in practice, it is applied at the exhaust, creating a forced draft in the dryer. Both cabinets and greenhouse dryers can be included in this category.

3.4.2 Indirect Active Solar Dryers

Indirect active solar dryers have a similar structure to indirect passive solar dryers. Here also a fan/blower can be used either at the inlet or outlet point of the dryer. However, in practice, the fan/blower is usually used at the inlet of the solar collector. A diagram sowing an indirect solar dryer is presented in Figure 3.15. Crops dried in this dryer are found to have good nutrient quality and color.

3.4.3 Mixed Mode Active Solar Dryers

The mixed mode active solar dryer has a similar structure to the mixed-mode passive solar dryer. There is only one difference, which is the incorporation of the fan/blower. A photograph of an example of a mixed mode active solar dryer, along with a schematic diagram, are presented in Figures 3.16 and 3.17. The blower can be powered by either a solar photovoltaic panel or electricity.

FIGURE 3.14
Photograph of a direct active solar dryer (Barnwal and Tiwari, 2008).

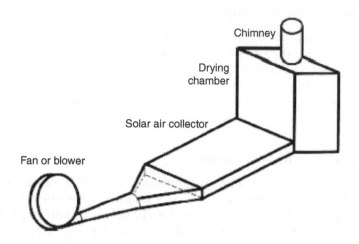

FIGURE 3.15
Schematic diagram of the indirect active solar dryer (Lingayat et al., 2020).

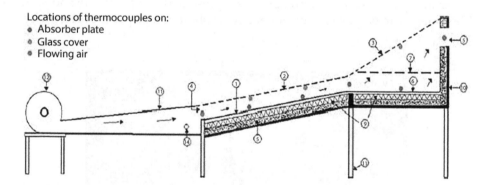

FIGURE 3.16
A schematic diagram of a mixed-mode active solar dryer: (1) collector absorber plate, (2) collector glass cover, (3) drying chamber glass cover, (4) inlet vent, (5) outlet vent, (6) drying chamber absorber plate, (7) wire mesh, (8) insulation, (9) electric heating plates, (10) wooden case, (11) angle iron stand, (12) blower, (13) divergent duct, and (14) small hole for measuring the air flow rate (Singh and Kumar, 2012).

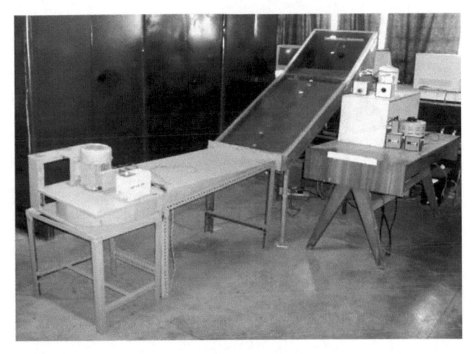

FIGURE 3.17
Photograph of a mixed-mode active solar dryer (Singh and Kumar, 2012).

Most of the researchers have run the blower using the solar photovoltaic panel to complete the process of being run by renewable energy.

3.5 Hybrid Solar Dryers

The hybrid solar dryer is the next step up from the active solar dryer. In such dryers, as well as solar energy, some other form of energy is involved. Sometimes biomass energy/electrical energy/heat exchange/desiccant is used to preheat the incoming air, and the heat transfer rate is enhanced inside the dryer. An example of a hybrid solar dryer is shown in Figures 3.18 and 3.19. Although hybrid dryers can be operated in both active and passive modes of heat transfer, the active mode is used in practice. In the active mode of the solar dryer, an exhaust fan is driven by a solar photovoltaic panel. It could also be run by grid-connected electricity, but in practice solar photovoltaic panels are used.

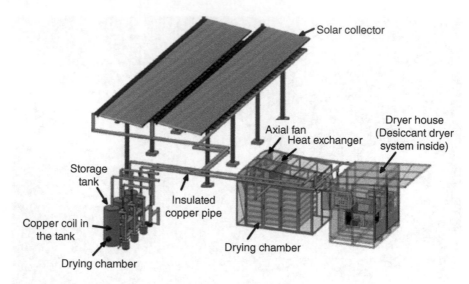

FIGURE 3.18
A schematic diagram of the hybrid solar dryer (Lamidi et al., 2019).

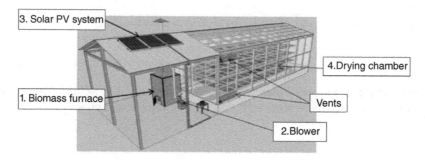

FIGURE 3.19
A schematic diagram of the biomass-based hybrid solar dryer (Lamidi et al., 2019).

3.6 Innovations in Solar Dryers

Solar dryers have attracted the attention of researchers working in many different fields, but all have a common interest, hence so much development has occurred with the classic solar dryer. Through these modifications, there has been the enhancement of drying efficiency and batch size, and drying times are now reduced.

A Subic-based company (a company working in the field of harvest loss: see http://www.sbma.com/news/2015/06/01/grainpro-launches-inno-vative-solar-dryer-to-help-small-farmers) has developed a solar bubble dryer 25 (SBD25), which offers enhanced grain drying capability. SBD25 is a 15 m long, collapsible modular dryer with a drying area of 25 m² that can accommodate up to 500 kg of grain. The SBD25 can be transported effort-lessly, assembled, and stored easily because of its compact design. This dryer can be used effectively by small-scale farmers producing 300–500 kilos per harvest.

S. Nabnean and colleagues (Nabnean et al., 2016) have developed a ther-mal storage assisted solar dryer, as shown in Figure 3.20. It was tested by drying osmotically dehydrated tomato cherries. Water is used as a thermal storage material, with 300 L of water being used for the purpose. The major components of this system are the thermal storage tank, solar air heater, heat exchanger, and drying chamber. The study reveals that the efficiency of the solar air heater is 21%–69%, and the payback period 1.37 years.

3.7 Miscellaneous Design Solar Dryer

The miscellaneous design solar dryer is a newly developed portable solar dryer. In this type of dryer, the solar collector is an integral part of the dry-ing chamber. The drying chamber is situated in either the roof or a wall. Figure 3.21 shows a miscellaneous solar dryer with roof integration.

This system consists of a solar collector assisted by a drying chamber inte-grated with the roof. The exhaust fan is operated by an electric motor to remove the inside air. The roof-integrated solar collector is covered with a polycarbonate sheet. The roof is painted black so that it acts as an absorber plate. The drying chamber has $1.3 \times 2.4 \times 0.8$ m³ capacity, and the building was sub-divided into two rooms.

An alternative version of an active solar dryer with solar roof collector is available. In this design, the solar collector is an integral part of the drying/ storage chamber's roof and a wall (Sahdev et al., 2017). In this system, air flows through the roof/wall collector via a duct. The moist air is transferred via an outlet vent situated on the other side of the wall.

The combined dryers use both solar energy and geothermal energy (Ivanova and Andonov, 2001). The drying chamber has 12 equally large trays where the crop is left to dry. The absorber plate is made of galvanized iron sheet (0.8 mm thick) painted black. The absorber surface has an effective area of 6 m². The absorber surface has an edge, which is sealed completely.

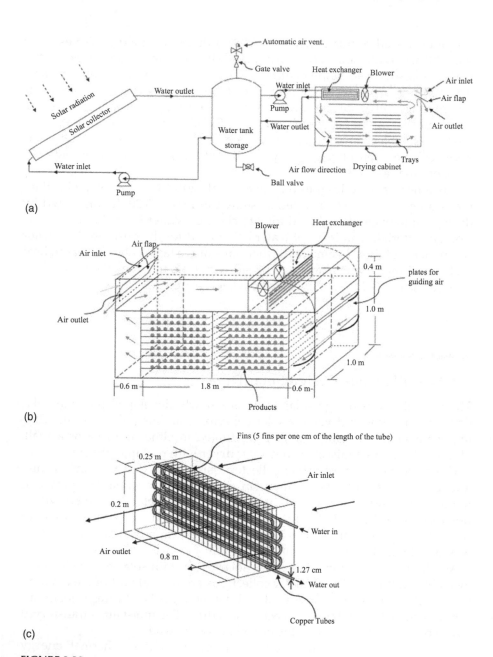

FIGURE 3.20
Schematic diagram of a solar dryer with water as thermal storage material: (a) system, (b) drying chamber, and (c) heat exchanger (Nabnean et al., 2016).

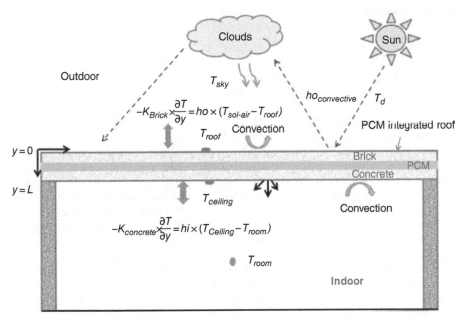

FIGURE 3.21
Miscellaneous roof integrated solar dryer (Bhamare et al., 2020).

The hot air moves into the empty geothermal water pipes and finally passes through the grating provided at the top of the pipes. The hot air then passes into the drying chamber. The dryer has a full load capacity of around 50 kg of fresh fruit and vegetables.

Problems

3.1 Describe the working principles of open sun drying, and direct and indirect solar drying methods.

3.2 Discuss the classification of solar dryers.

3.3 Differentiate between passive and active solar dryers.

3.4 Discuss various designs of passive solar dryers.

3.5 Discuss different types of active solar dryers.

3.6 What are mixed-mode solar dryers? Discuss briefly.

3.7 What are solar greenhouse dryers?

3.8 Discuss various designs of hybrid photovoltaic-solar dryers.

References

Abubakar, S., Umaru, S., Kaisan, M.U., Umar, U.A., Ashok, B. and Nanthagopal, K., 2018. Development and performance comparison of mixed-mode solar crop dryers with and without thermal storage. *Renewable Energy*, 128, 285–298.

Agrawal, A. and Sarviya, R.M., 2016. A review of research and development work on solar dryers with heat storage. *International Journal of Sustainable Energy*, 35(6), 583–605.

Akpinar, E.K., 2006. Mathematical modelling of thin layer drying process under open sun of some aromatic plants. *Journal of Food Engineering*, 77(4), 864–870.

Barnwal, P. and Tiwari, G.N., 2008. Grape drying by using hybrid photovoltaic-thermal (PV/T) greenhouse dryer: an experimental study. *Solar energy*, 82(12), 1131–1144.

Bhamare, D.K., Rathod, M.K. and Banerjee, J., 2020. Numerical model for evaluating thermal performance of residential building roof integrated with inclined phase change material (PCM) layer. *Journal of Building Engineering*, 28, 101018.

César, L.V.E., Lilia, C.M.A., Octavio, G.V., Isaac, P.F. and Rogelio, B.O., 2020. Thermal performance of a passive, mixed-type solar dryer for tomato slices (*Solanum lycopersicum*). *Renewable Energy*, 147.845–855.

Chauhan, P.S., Kumar, A. and Gupta, B., 2017. A review on thermal models for greenhouse dryers. *Renewable and Sustainable Energy Reviews*, 75, 548–558.

Chauhan, P.S. and Kumar, A., 2018. Thermal modeling and drying kinetics of gooseberry drying inside north wall insulated greenhouse dryer. *Applied Thermal Engineering*, 130, 587–597.

Essalhi, H., Benchrifa, M., Tadili, R. and Bargach, M.N., 2018. Experimental and theoretical analysis of drying grapes under an indirect solar dryer and in open sun. *Innovative Food Science & Emerging Technologies*, 49, 58–64.

Huddar, V.B. and Kamoji, M.A., 2019, March. Experimental investigation on performance of small passive solar greenhouse dryer for cashew kernel drying. In *AIP Conference Proceedings* (Vol. 2080, No. 1, p. 030001). AIP Publishing, Karnataka, India.

Iranmesh, M., Akhijahani, H.S. and Jahromi, M.S.B., 2020. CFD modeling and evaluation the performance of a solar cabinet dryer equipped with evacuated tube solar collector and thermal storage system. *Renewable Energy*, 145, 1192–1213.

Ivanova, D. and Andonov, K., 2001. Analytical and experimental study of combined fruit and vegetable dryer. *Energy Conversion and Management*, 42(8), 975–983.

Jain, D. and Tiwari, G.N., 2004. Effect of greenhouse on crop drying under natural and forced convection II. Thermal modeling and experimental validation. *Energy Conversion and Management*, 45(17), 2777–2793.

Kannan, N. and Vakeesan, D., 2016. Solar energy for future world: A review. *Renewable and Sustainable Energy Reviews*, 62, 1092–1105.

Kumar, A. and Tiwari, G.N., 2006. Thermal modeling of a natural convection greenhouse drying system for jaggery: An experimental validation. *Solar Energy*, 80(9), 1135–1144.

Lakshmi, D.V.N., Layek, A. and Muthukumar, P., 2018, October. Evaluation of Convective Heat Transfer Coefficient of Herbs Dried in a Mixed Mode Solar Dryer. In *2018 International Conference and Utility Exhibition on Green Energy for Sustainable Development (ICUE)* (pp. 1–7). IEEE, Phuket, Thailand.

Lamidi, R.O., Jiang, L., Pathare, P.B., Wang, Y. and Roskilly, A.P., 2019. Recent advances in sustainable drying of agricultural produce: A review. *Applied Energy*, 233, 367–385.

Lingayat, A.B., Chandramohan, V.P., Raju, V.R.K. and Meda, V., 2020. A review on indirect type solar dryers for agricultural crops–Dryer setup, its performance, energy storage and important highlights. *Applied Energy*, 258, 114005.

Madhlopa, A., Jones, S.A. and Saka, J.K., 2002. A solar air heater with composite–absorber systems for food dehydration. *Renewable Energy*, 27(1), 27–37.

Motahayyer, M., Arabhosseini, A. and Samimi-Akhijahani, H., 2019. Numerical analysis of thermal performance of a solar dryer and validated with experimental and thermo-graphical data. *Solar Energy*, 193, 692–705.

Nabnean, S., Janjai, S., Thepa, S., Sudaprasert, K., Songprakorp, R. and Bala, B.K., 2016. Experimental performance of a new design of solar dryer for drying osmotically dehydrated cherry tomatoes. *Renewable Energy*, 94, 147–156.

Pandey, A.K., Hossain, M.S., Tyagi, V.V., Rahim, N.A., Jeyraj, A., Selvaraj, L. and Sari, A., 2018. Novel approaches and recent developments on potential applications of phase change materials in solar energy. *Renewable and Sustainable Energy Reviews*, 82, 281–323.

Prakash, O., Kumar, A. and Laguri, V., 2016a. Performance of modified greenhouse dryer with thermal energy storage. *Energy Reports*, 2, 155–162.

Prakash, O., Laguri, V., Pandey, A., Kumar, A. and Kumar, A., 2016b. Review on various modelling techniques for the solar dryers. *Renewable and Sustainable Energy Reviews*, 62, 396–417.

Sahdev, R.K., Kumar, M. and Dhingra, A.K., 2017. Effect of mass on convective heat transfer coefficient during open sun drying of groundnut. *Journal of Food Science and Technology*, 54(13), 4510–4516.

Sahu, T.K., Gupta, V. and Singh, A.K., 2016. A review on solar drying techniques and solar greenhouse dryer. *IOSR Journal of Mechanical and Civil Engineering (IOSR-JMCE)*, 13(3), 31–37.

Selvaraj, M., Sadagopan, P., Balakrishnan, N. and Bhuvaneswaran, M., 2018. A Review of solar energy collection technology to heat air as thermal, using flat plate collector and integrated with drying chamber for drying food products. *Indian Journal of Scientific Research*, 281–287.

Sharma, A., Chen, C.R. and VuL an, N., 2009. Solar-energy drying systems: A review. *Renewable and Sustainable Energy Reviews* 13(6–7), 1185–1210.

Singh, S. and Kumar, S., 2012. Testing method for thermal performance based rating of various solar dryer designs. *Solar Energy* 86(1), 87–98.

Slimani, M.E.A., Amirat, M., Bahria, S., Kurucz, I. and Sellami, R., 2016. Study and modeling of energy performance of a hybrid photovoltaic/thermal solar collector: Configuration suitable for an indirect solar dryer. *Energy Conversion and Management*, 125, 209–221.

Sodha, M.S., Dang, A., Bansal, P.K. and Sharman, S.B., 1985. An analytical and experimental study of open sun drying and a cabinet tyre drier. *Energy Conversion and Management*, 25(3), 263–271.

Yadav, S., Lingayat, A.B., Chandramohan, V.P. and Raju, V.R.K., 2018. Numerical analysis on thermal energy storage device to improve the drying time of indirect type solar dryer. *Heat and Mass Transfer*, 54(12), 3631–3646.

4

Performance Analysis of Solar Drying Systems

4.1 Introduction

The performance analysis of dryers is very important. By using proper performance analysis, farmers/entrepreneurs/industrialists can gain maximum advantage. This will not only increase revenue but also employability, along with strengthening the national economy. This analysis of solar drying systems has also become important because it depends on ambient parameters, and these ambient parameters are dependent on climatic conditions. Climatic conditions vary with changes in angles of latitude and longitude, and these angles vary from place to place (Fudholi et al., 2012).

Since the 1990s, intensive research has been undertaken in the field of solar drying, with the main aim being to enhance the method's effectiveness. This has led to the development of various performance parameters. At present, all parameters for the comparison of solar dryers are based on either the first or second laws of thermodynamics. Based on the first law, these are important performance parameters such as drying efficiency, heat utilization factor, coefficient of performance, thermal efficiency, and overall daily thermal efficiency. However, based on the second law, exergy efficiency is the most important performance parameter. Researchers have validated innovative solar dryers, or modifications to existing dryers, by comparing their performance to existing dryers based on these parameters only.

In order to discover the complete thermal profile of the solar dryer, the dryer is first tested under no-load conditions. The advantage of this is that there is no interference with performance through external factors. In load conditions, the inner temperature or drying room temperature and relative humidity are affected by the crop temperature and its moisture content. The testing of solar dryers in no-load conditions depends mainly on ambient parameters and the construction of the dryers themselves. Hence it is very important to carry out no-load testing of newly developed or modified solar dryers. Various researchers have proposed testing procedures for solar dryers in the no-load condition.

Leon et al. (2002) presented a detail testing procedure for solar dryers. Separate testing procedures are adopted in no-load and load conditions. Prakash and Kumar (2014) designed and developed a modified greenhouse dryer. The system was a modified version of the classic even-span-roof type of greenhouse dryer, using a natural convection mode of heat transfer. The modification was applied to the north wall of the system and the ground area of the dryer. The system was tested in the no-load condition. Sutar and Tiwari (1996) had developed and presented a characteristic curve for greenhouse dryers. This curve should intercept at zero in an efficient solar greenhouse dryer system. Results show that the characteristic curve of the modified greenhouse dryer also intercepted at zero, thus justifying the modification made in the greenhouse dryer.

4.2 Drying Efficiency

Drying efficiency is the most important parameter used to analyze the performance of the dryer. It is the ratio between heat utilized per total heat consumed during the process of drying. Mathematically it is expressed as follows:

$$\eta_s = \frac{Q_a}{Q_c} \tag{4.1}$$

Where Q_a is the heat utilized in kJ and Q_c is the heat consumed in kJ.

For natural convection solar dryers

In this type of dryer, heat transfer takes place via the thermosphyon effect, which leads to the greenhouse effect.

$$\eta_s = \frac{WL}{IA} \tag{4.2}$$

Where W is weight of removal of water content in kg, L is the latent heat of vaporization in kJ/kg, I is the solar intensity in W/m^2, and A is the collector area in m^2.

For dryers operated under the forced convection mode of heat transfer

In this case, the forced convection is being generated by a blower or fan. Hence the total consumed energy is increased because of the additional energy consumed by running the blower. The mathematical expression of the drying efficiency in this condition is given below:

$$\eta_s = \frac{WL}{IA + p_f} \tag{4.3}$$

Where P_f is the energy consumed to run the blower. The unit is in W.
For hybrid solar dryers

These types of solar dryer are more effective because they utilize both types of energy—conventional and non-conventional—but sometimes only non- conventional energy is used. However, it utilizes more than one concept. Sometime hybrid solar dryers use LPG or biogas to enhance the drying efficiency of the system. The drying efficiency of the hybrid solar dryer is expressed as follows:

$$\eta_s = \frac{WL}{(IA + p_f) + (m_b \times LCV)} \tag{4.4}$$

Where m_b is the mass of biomass in kg and LCV is the lower calorific value in kJ/kg.

4.3 Coefficient of Performance

The coefficient of performance is the ratio between the net rise of the drying room temperature compared to the ambient temperature per net rise of the bed temperature as compared to the ambient temperature. Since it is a ratio, this is a unitless and dimensionless parameter. Mathematically, it is expressed as follows:

$$COP = \frac{T_r - T_a}{T_f - T_a} \tag{4.5}$$

4.4 Thermal Efficiency

The thermal efficiency of a solar dryer is the ratio of heat utilized in the evaporation of moisture from the product per total heat supplied by thermal energy (Luh and Woodroof, 1975). Mathematically, it is expressed as follows:

$$\eta_c = \frac{mc(T_o - T_i)}{A_c S} \times 100 \tag{4.6}$$

where,

m = mass flow rate (kg/s)
C = specific heat of air (J kg⁻¹°C⁻¹)
A_c = collector area (m²)
T_i = inlet air temperature (°C)
T_o = outlet air temperature (°C)
S = solar radiation intensity (W/m²)

4.5 Overall Daily Thermal Efficiency

Overall daily thermal efficiency is the thermal efficiency calculated on a daily basis. It uses the same mathematical expression as for the thermal efficiency of the dryer. However, the time frame of the computation is taken for one day only. Therefore, overall daily thermal efficiency is not a constant but varies on a day-to-day basis. In general, the first day of drying has a higher overall daily thermal efficiency compared to the other days. As the days pass, the overall daily thermal efficiency decreases.

4.6 Exergy Analysis

Thermal efficiency assessed via exergy analysis is based on the second law of thermodynamics, and is associated with mass, heat, and work transfer in the specified system. This analysis is used to forecast the thermal behavior of the system. Exergy analysis has a wide range of applications, and this concept is applicable to the solar drying process.

These are the imortant concepts n the exergy analysis of solar drying systems. They are also based on the first law of energy balance for the system. In a steady state condition, the exergy analysis of the system is expressed as follows (Akbulut and Durmus, 2010):

$$Ex = m_{da}c_{pda}\left[(T - T_a) - T_a \ln \frac{T}{T_a}\right] \tag{4.7}$$

The exergy analysis at the inlet of the solar dryer is expressed as follows:

$$Ex_{dci} = m_{da}c_{pda}\left[(T_{dci} - T_a) - T_a \ln \frac{T_{dci}}{T_a}\right] \tag{4.8}$$

The exergy analysis at the outlet of the solar dryer is expressed as follows:

$$Ex_{dco} = m_{da}c_{pda}\left[\left(T_{dco} - T_a\right) - T_a \ln\frac{T_{dco}}{T_a}\right] \tag{4.9}$$

The net exergy loss during the process of solar drying is expressed as follows:

$$Ex_{loss} = Ex_{dci} - Ex_{dco} \tag{4.10}$$

The exergy efficiency is the ratio of the outlet exergy per inlet exergy of the solar dryer. Mathematically, it is expressed as follows (Akbulut and Durmus, 2010; Akpinar, 2010; Fudholi et al., 2014a):

$$\eta_{Ex} = \frac{Ex_{dco}}{Ex_{dci}} = 1 - \frac{Ex_{loss}}{Ex_{dco}} \tag{4.11}$$

4.7 Exergy Analysis of Various Solar Dryers

Fudholi et al. (2014a) carried out exergy analysis of solar dryers (as shown in Figure 4.1).

The system was used to dry red chilis. The distribution of the exergy balance is shown in Figure 4.2.

The performance analysis of the dryer is presented in Table 4.1.

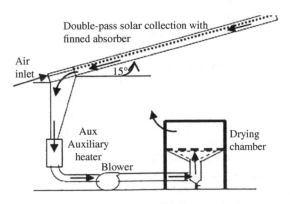

FIGURE 4.1
Schematic diagram of the indirect solar drying system.

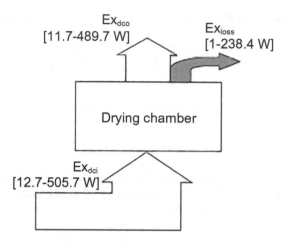

FIGURE 4.2
Schematic diagram of the exergy flow of the system, where Ex_{dci} is the inlet of the exergy in W, Ex_{dco} is the outlet of exergy in W, and Ex_{loss} is the exergy loss in W.

TABLE 4.1

Exergy Analysis of the System

Parameters	Unit	Values
Initial weight of product (total)	kg	40
Final weight of product (total)	kg	8
Initial moisture content (wet basis)	%	80
Final moisture content (wet basis)	%	10
Air mass flow rate	kg/s	0.07
Average solar radiation	W/m²	420
Average ambient temperature	°C	30
Average drying chamber temperature	°C	44
Average ambient relative humidity	%	62
Average drying chamber humidity	%	33
Drying time	h	33
Blower energy	kWh	4.13
Solar energy	kWh	160.43
Evaporative capacity	kg/h	0.97
Specific energy consumption	kWh/kg	5.26
Overall heat collection (thermal) efficiency	%	28
Overall drying efficiency, up to 10% wb	%	13
Pick-up efficiency, up to 10% wb	%	45
Overall exergy efficiency, up to 10% wb	%	57
Overall improvement potential, up to 10% wb	W	47.29

Tiwari and Tiwari (2017) had developed a mixed-mode greenhouse-type solar dryer with an additional N-PVT air collector for the purpose of energy and exergy analysis. The developed system is shown in Figure 4.3.

For the given system, exergy analysis is conducted, exergy efficiency and equivalent exergy efficiency are evaluated across all five conditions of the system. The result is presented in Figure 4.4. It varies between 61.56% and 42.22%, and 28.96% and 19.11%, respectively.

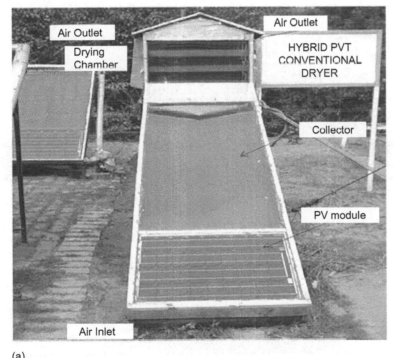

(a)

FIGURE 4.3
(a) Photograph of the system.   (*Continued*)

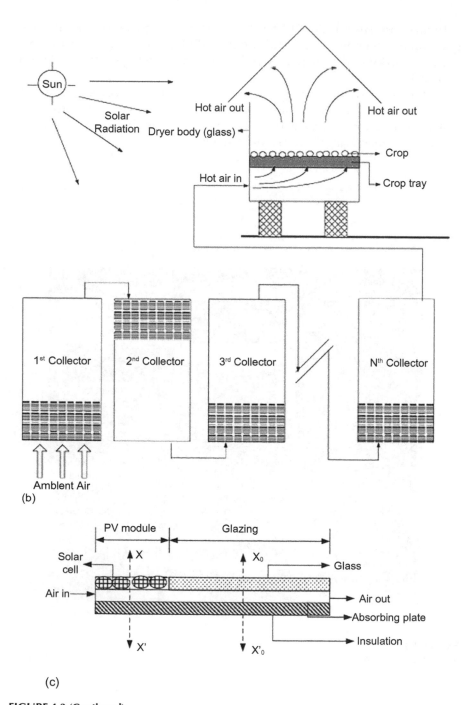

FIGURE 4.3 (Continued)
(b) schematic diagram of the system, and (c) cross-sectional view of the used air collector.

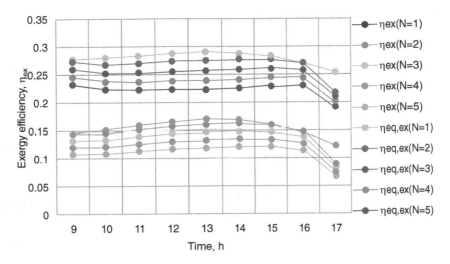

FIGURE 4.4
Hourly variation of the exergy efficiency and equivalent exergy efficiency.

4.8 Testing Procedure of Solar Dryers in Load and No-Load Conditions

4.8.1 Load Condition

Singh and Kumar (2012a) established a dimensionless parameter to rate the timber drying operation in the loaded state. It is termed the drier operation indicator (DPI)

$$\phi = \exp(-DPI \times k \times t) \tag{4.12}$$

Where t is drying time in seconds and k is the drying constant, ϕ is the moisture content of the agricultural produce. DPI is a dimensionless parameter.

4.8.2 No-Load Condition

Singh and Kumar (2012b) have developed the dimensionless parameter NLPI (No load performance index) for solar dryers. The equation relates to the steady state heat flow condition. Mathematically, it is expressed as follows:

$$\frac{m_1 C_p}{A_s h_{cpfs}} = \frac{T_{mp} - T_{mf}}{T_{mfo} - T_{mfi}} \left[\frac{I_g - U_b\left(T_{mp} - T_{am}\right) - U_t\left(T_{mg} - T_{am}\right)}{I_g - U_b\left(T_{mp} - T_{am}\right) - h_{rpg}\left(T_{mp} - T_{mg}\right)} \right] = NLPI \tag{4.13}$$

The required unknown parameters are being calculated as:

$$h_{rpg} = \frac{\sigma\left[\left(T_{mp}+273\right)^2+\left(T_{mg}+273\right)^2\right]\left[\left(T_{mp}+273\right)+\left(T_{mg}+273\right)\right]}{\left[\dfrac{1}{\varepsilon_p}+\dfrac{1}{\varepsilon_g}-1\right]} \tag{4.14}$$

$$h_{cpf} = \frac{S-U_b\left(T_{mp}-T_{am}\right)-h_{rpg}\left(T_{mp}-T_{mg}\right)}{\left(T_{mp}-T_{mf}\right)} \tag{4.15}$$

$$U_b = \frac{K_b}{L_b} \tag{4.16}$$

$$U_t = h_{rga}+h_{cga} \tag{4.17}$$

$$h_{cga} = \frac{\sigma\varepsilon_g\left[\left(T_{mg}+273\right)^4-\left(T_{am}+273\right)^4\right]}{\left(T_{mg}-T_{am}\right)} \tag{4.18}$$

$$h_{cga} = 0.54\left(Ra\right)^{0.25}\left[\frac{K}{L}\right] \tag{4.19}$$

Where there are nomenclatures in the equations:

A_s is the surface area of the absorber plate of the dryer system (m²)

h_{cpfs} is the convection heat transfer coefficient between the absorber plate and flowing inside air (W/m²K)

U_b is the heat transfer loss coefficient at the bottom (W/m²K)

h_{rpg} is the radiative heat transfer coefficient between the absorber plate and glass cover

T_{mp} is the mean absorber plate temperature (°C)

T_{mf} is the mean inside flowing air temperature (°C)

T_{mfo} is the mean outgoing flowing air temperature (°C)

T_{mfi} is the mean incoming flowing air temperature to the collector (°C)

I_g is the solar radiation (W/m²)

T_{am} is the ambient temperature (°C)

T_{mg} is the mean glass temperature (°C)

ε_p is the emissivity of the absorber plate (which is taken as 0.9)

ε_g is the emissivity of the glass cover (which is taken as 0.9)

K_b is the conductivity of the insulation

L_b is the thickness of the insulation

m_1 is the air flow rate (kg/s)

C_p is the specific heat of air (J/kgK)

4.9 Discussion

Singh and Kumar (2012b) have developed a systematic procedure to evaluate the thermal performance of solar dryers in the no-load situation. They have named one performance parameter as the No-Load Performance Index (NLPI). This concept is applied to all three types of solar dryer, namely direct, indirect, and mixed-mode solar dryers. The schematic diagram and a photograph of the cabinet/direct solar dryer is shown in Figure 4.5.

Table 4.2 shows in detail the evaluation of the NLPI for the cabinet dryer in passive/natural convection heat transfer mode.

Table 4.3 shows the detailed presentation of NLPI with experimental data for the indirect solar dryer.

Table 4.4 shows the detailed computation of NLPI for the mixed-mode solar dryer with the help of experimental data.

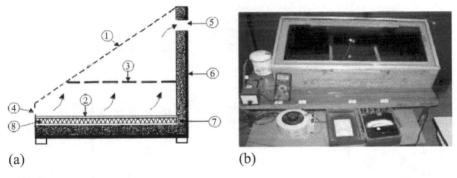

(a)   (b)

FIGURE 4.5
(a) Schematic diagram of the cabinet dryer under passive mode, where 1 is the glass cover, 2 absorber plate, 3 wire mesh tray, 4 inlet holes, 5 outlet holes, 6 wooden case, 7 insulation, and 8 electric heating plate; (b) photograph of a model of the dryer.

TABLE 4.2

Detailed Presentation of NLPI for Cabinet/Direct Dryer Under Passive Mode

Absorbed Thermal Energy (W/m²)	Plate Temp. (°C)	Hot Air, Temp. (°C)	Glass Cover, Temp (°C)	Ambient Air, Temp. (°C)	No-load Performance Index (NLPI)
286	46.94	24.82	30.32	14.67	0.385
403	55.75	26.77	33.50	13.94	0.441
504	64.30	29.95	36.68	14.67	0.465
571	70.17	32.03	40.34	14.67	0.433
588	77.02	38.51	46.46	20.78	0.425
605	71.64	32.52	39.85	14.67	0.452
672	77.75	34.84	42.79	15.40	0.461
756	89.98	42.79	54.52	22.01	0.453
806	86.06	36.43	46.46	13.94	0.453

TABLE 4.3

Detailed Presentation of NLPI for the Indirect Solar Dryer

Absorbed Thermal Energy (W/m2)	Ambient Air Temp (°C)	Air Collector			Dryer			No-load Performance Index (NLPI)
		Plate Temp. (°C)	Hot Air, Temp. (°C)	Glass Cover, Temp (°C)	Plate Temp. (°C)	Hot Air, Temp. (°C)	Glass Cover, Temp (°C)	
1. Natural convection mode								
300	24.69	53.91	35.57	34.96	44.01	44.62	39.85	0.483
450	24.94	64.67	38.51	36.80	45.23	49.88	42.30	0.476
550	24.69	68.34	38.88	37.04	48.41	51.59	43.28	0.523
600	25.43	71.27	40.22	38.14	50.12	53.18	44.25	0.546
650	25.43	74.21	41.81	39.12	53.79	55.99	48.17	0.541
700	25.43	77.38	43.03	41.69	56.24	57.99	48.66	0.553
750	24.94	79.46	44.25	43.89	61.13	61.13	53.06	0.519
800	26.16	82.40	46.09	44.99	65.04	63.33	54.52	0.553
2. Forced convection mode								
(a) Mass flow rate: 0.009 kg/s								
300	29.34	54.16	36.43	35.82	42.05	41.07	39.12	0.918
450	30.07	62.47	38.26	37.65	45.48	44.98	44.01	0.9
600	30.07	69.07	39.85	38.88	48.66	48.04	46.94	0.899
750	29.34	75.06	40.71	40.10	50.86	49.62	48.17	0.959
800	30.07	78.73	42.54	41.69	53.30	52.06	50.61	0.939

(Continued)

TABLE 4.3 (Continued)

Detailed Presentation of NLPI for the Indirect Solar Dryer

Absorbed Thermal Energy (W/m2)	Ambient Air Temp (°C)	Air Collector			Dryer			No-load Performance Index (NLPI)
		Plate Temp. (°C)	Hot Air, Temp. (°C)	Glass Cover, Temp (°C)	Plate Temp. (°C)	Hot Air, Temp. (°C)	Glass Cover, Temp (°C)	
(b) Mass flow rate: 0.017 kg/s								
300	29.34	47.19	32.64	31.79	36.19	35.32	34.72	1.323
450	30.07	52.32	33.13	32.76	37.41	36.67	36.19	1.331
600	30.07	57.95	34.23	34.23	39.12	38.37	37.65	1.375
750	29.34	63.94	36.06	36.06	42.30	41.17	40.59	1.388
810	30.07	65.65	37.65	37.65	46.46	43.38	42.30	1.349
(c) Mass flow rate: 0.026 kg/s								
300	27.63	43.15	30.93	32.15	34.96	33	32.27	1.432
450	26.90	47.31	31.54	30.93	34.72	35.72	32.76	1.432
600	27.87	51.96	32.52	32.03	35.70	35.30	33.74	1.486
750	27.87	55.50	33.25	32.52	38.63	36.67	35.21	1.518
800	30.07	59.78	36.31	35.70	40.634	39.36	36.19	1.509

TABLE 4.4

Detailed Presentation of NLPI for the Mixed-Mode Solar Dryer

Absorbed Thermal Energy (W/m²)	Ambient Air Temp. (°C)	Air collector			Dryer			No-Load Performance Index (NLPI)
		Plate Temp. (°C)	Hot Air, Temp. (°C)	Glass Cover, Temp. (°C)	Plate Temp. (°C)	Hot Air, Temp. (°C)	Glass Cover, Temp. (°C)	
1. Natural convection mode								
300	24.45	53.30	34.35	32.15	59.41	45.72	36.19	0.554
375	24.45	59.17	36.31	33.37	66.02	50.12	39.61	0.538
450	24.45	62.47	37.78	33.74	73.35	53.79	42.79	0.543
550	24.69	70.78	40.34	36.06	80.69	59.41	46.46	0.635
650	24.69	76.90	42.79	37.90	89.24	64.79	51.35	0.611
700	25.18	80.32	44.38	39.49	94.13	68.22	52.81	0.607
750	24.69	82.89	45.84	39.49	97.80	68.95	54.28	0.632
800	26.65	86.68	48.17	40.34	102.69	72.13	59.90	0.641
2. Forced convection								
(a) Mass flow rate: 0.009 kg/s								
250	33.01	54.89	38.75	37.65	59.41	45.97	41.08	1.053
300	33.01	56.60	39.73	38.88	62.59	46.94	42.05	1.062
450	31.05	61.49	41.93	40.10	73.11	48.90	44.74	1.086
600	31.79	72.25	46.93	42.30	84.60	55.01	48.90	1.088
750	31.05	77.87	46.70	43.64	93.40	58.68	51.10	1.094
800	32.27	81.30	49.02	44.01	99.51	62.35	53.06	1.085

(Continued)

TABLE 4.4 (Continued)

Detailed Presentation of NLPI for the Mixed-Mode Solar Dryer

Absorbed Thermal Energy (W/m²)	Ambient Air Temp. (°C)	Air collector			Dryer			No-Load Performance Index (NLPI)
		Plate Temp. (°C)	Hot Air, Temp. (°C)	Glass Cover, Temp. (°C)	Plate Temp. (°C)	Hot Air, Temp. (°C)	Glass Cover, Temp. (°C)	
(b) Mass flow rate: 0.017 kg/s								
300	32.76	54.40	36.31	34.72	51.35	42.05	38.88	1.478
450	33.74	57.46	37.65	36.92	64.30	46.46	41.57	1.486
600	32.52	62.35	39.24	37.90	69.44	47.19	42.30	1.527
750	33.50	67.97	41.20	39.24	76.77	50.61	45.23	1.534
800	34.72	70.66	42.42	40.71	80.93	53.79	46.46	1.451
(c) Mass flow rate: 0.026 kg/s								
250	29.34	40.83	31.91	30.93	45.97	35.45	33.01	1.604
300	33.01	44.62	34.47	33.01	52.57	40.34	36.68	1.603
450	32.27	51.83	35.45	33.99	55.01	41.57	37.90	1.633
600	31.30	57.34	36.68	35.09	59.66	42.79	39.61	1.619
750	32.76	61.25	38.63	36.19	66.99	46.21	42.30	1.616
800	34.47	65.40	40.46	39.98	70.42	49.39	43.52	1.536

By doing this comparative analysis, the mixed-mode solar dryer shows a higher value of NLPI compared to results for cabinet and indirect solar dryers.

4.10 Performance Characteristics of Various Solar Dryers

The following are important characteristics for solar dryers, namely:

ii. The drying atmosphere, such as the air temperature, relative humidity, and air flow speed.

iii. Product factors such as product shape, size, thickness, amount, initial and moisture content.

iiii. The dryers' constructional factors, such as shape, size, drying tray area, etc.

These are important parameters to be noted and quantified:

- Type, size, shape
- Sensory quality
- Physical features of the dryer
- Tray area and number
- Drying capacity/loading density
- Dryer efficiency
- Convenient loading and unloading
- Nutritional attributes
- Thermal performance
- Rehydration capacity
- Drying time and rate
- Airflow rate
- Drying air temperature and relative humidity
- Quality of dried product
- Cost of dryer and payback period

Problems

4.1 Explain the significance of the performance parameters for solar dryers

4.2 Discuss the importance of the exergy analysis of solar dryers.

4.3 Discuss the various parameters for the performance evaluation of solar dryers based on the first and second laws of thermodynamics.

References

Akbulut, A. and Durmus, A., 2010. Energy and exergy analyses of thin layer drying of mulberry in a forced solar dryer. *Energy* 35, 1754–1763.

Akpinar, E.K., 2010. Drying of mint leaves in a solar dryer and under open sun: Modelling, performance analyses. *Energy Conversion Management* 51, 2407–2418.

Fudholi, A., Ruslan, M.H., Othman, M.Y., Azmi, M.S.M., Zaharim, A. and Sopian, K., 2012. Drying of palm oil fronds in solar dryer with finned double-pass solar collectors. *WSEAS Transactions on Heat and Mass Transfer* 4(7), 105–114.

Fudholi, A., Sopian, K., Othman, M.Y. and Ruslan, M.H., 2014a. Energy and exergy analyses of solar drying system of red seaweed. *Energy Buildings* 68, 121–129.

Fudholi, A., Sopian, K., Yazdi, M.H., Ruslan, M.H., Gabbasa, M. and Kazem, H.A., 2014b. Performance analysis of solar drying system for red chili. *Solar Energy* 99, 47–54.

Leon, M.A., Kumar, S., and Bhattacharya, S.C., 2002. A comprehensive procedure for performance evaluation of solar food dryers. *Renewable and Sustainable Energy Reviews*, 6, 367–393.

Luh, B.S. and Woodroof, J.G., 1975. *Commercial vegetable processing*. USA: The Avi Publishing Company.

Prakash, O. and Kumar, A., 2014. Design, development, and testing of a modified greenhouse dryer under conditions of natural convection. *Heat Transfer Research* 45(5), 433–451

Singh, S. and Kumar, S., 2012a. New approach for thermal testing of solar dryer: Development of generalized drying characteristic curve. *Solar Energy* 86(7), 1981–1991.

Singh, S. and Kumar, S., 2012b. Testing method for thermal performance based rating of various solar dryer designs. *Solar Energy* 86(1), 87–98.

Sutar, R.F. and Tiwari, G.N., 1996. Temperature reductions inside a greenhouse. *Energy* 21(1), 61–65.

Tiwari, S. and Tiwari, G.N., 2017. Energy and exergy analysis of a mixed-mode greenhouse-type solar dryer, integrated with partially covered N-PVT air collector. *Energy* 128, 183–195.

5

Thermal Modeling of Solar Drying Systems

5.1 Introduction

The solar drying of hygroscopic agricultural produce is a complex process of heat and mass transfer (Ahmad and Prakash, 2020). Hence, in order to improve drying efficiency and the drying rate, thermal modelling plays an important role and is necessary in the design and development of effective solar dryers. It optimizes the important parameters to take maximum advantage of the available resources under various operating conditions (Chauhan et al., 2018). The important parameters for thermal modelling are inside air temperature, inside air relative humidity, crop weight, crop temperature, ambient temperature, and ambient parameters (Arun et al., 2019). With the help of thermal modelling, inside air temperature, inside air relative humidity drying rate, drying kinetics and drying potentials can be predicted with great accuracy. The convective heat transfer coefficient is also an important parameter for the thermal modelling of solar dryers.

Kumar and Tiwari (2006a) had carried out thermal modelling for greenhouse dryers in the forced convection mode. The system was used to test conditions for drying jaggery, a product made from concentrated sugar cane juice. The study was conducted with the size of the jaggery blocks being 0.03 × 0.03 ×0.03 m³ and each weighing 2.0 kg. The inside floor area of the dryer was to be 0.936 m². The experimental observation was conducted in March 2004 at IIT Delhi. The experiment lasted from 10 am to 5 pm. The modelling was programmed with the aid of MATLab software. The thermal model was used to predict three important parameters: crop temperature, room temperature, and crop moisture evaporation. The prediction model was validated with experimental observation.

5.2 Convective Heat and Mass Transfer

5.2.1 Convective Heat Transfer

Convective heat transfer is also known as the convective mode of heat transfer, where heat transfer takes place from flowing air to a solid surface. This mode of heat transfer is very prominent in solar dryers. The thermal model used the convective heat transfer coefficient (W/m^{2}K) for its predictions. Prakash and Kumar (2014a) developed a modified greenhouse dryer under a natural convection mode. In this system, two major modifications were incorporated into a conventional even-span-roof type of greenhouse dryer, namely to the north wall and the floor of the dryer. The dryer was tested under two different floor conditions, namely a bare concrete floor, and a concrete floor covered with black PVC sheeting. With the experimental data set to no-load condition, the convective heat transfer coefficient was computed. A comparative study was conducted with reference to the convective heat transfer coefficient when the dryer is operated with these two floor conditions. The study revealed that the value of the convective heat transfer coefficient for the dryer operating over a floor covered with black PVC sheeting is higher than for a bare concrete floor. The floor covered with black PVC sheet shows a high heat absorption capacity and less heat loss as compared to the bare concrete floor.

Prakash and Kumar (2014b) studied a modified greenhouse dryer under active/forced convection mode. The study was conducted in the no-load condition and with two different floor conditions, namely a bare floor and a floor covered with black PVC sheeting. To create forced convection, an exhaust fan was applied at the outlet vent, powered by a polycrystalline cell. Study revealed that the dryer operating with the floor covered with black PVC sheeting showed a higher convective heat transfer coefficient compared to the result with the bare floor.

Kumar and Tiwari (2006b) developed a thermal model for a greenhouse dryer under natural convection mode. This greenhouse dryer has the even-span type of roof, as shown in Figure 5.1.

The experimental observations are presented in Table 5.1. All input parameters for the thermal model are presented in this Table. There are three parameters used for prediction by this model, namely crop temperature, crop moisture evaporation, and dryer room temperature. All three predicted parameters were validated with experimental data. Statistical analysis was applied to the thermal model.

The thermal energy flow of the dryer is presented in Figure 5.2.

The experimental data for this is presented in Table 5.1.

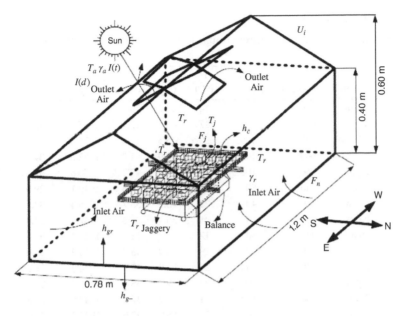

FIGURE 5.1
Schematic diagram of the even-span-roof type of greenhouse dryer.

TABLE 5.1

Experimental Observation of a Greenhouse Dryer for Jaggery Drying Under the Passive Mode

Day Drying	Time	$I(t)$ (W/m²)	$I(d)$ (W/m²)	T_a (°C)	T_j (°C)	T_e (°C)	$Mj \times 10^{-3}$ kg	γ_r (%)	γ_a (%)
1	10 am	340	80	14	38.5	30.5	2000.0	55.2	56.0
	11 am	520	80	18	44.2	33.8	1993.2	45.8	45.4
	12 noon	560	100	19	54.7	43.5	1987.0	37.7	36.1
	1 pm	480	100	20	55.6	45.9	1952.4	35.6	35.1
	2 pm	560	100	23	58.6	49.3	1978.5	34.0	33.8
	3 pm	360	80	23	54.9	46.2	1975.5	33.1	32.3
	4 pm	220	60	21	50.6	43.3	1973.6	34.2	33.5
	5 pm	60	20	20	45.5	40.3	1971.9	38.9	38.3
2	10 am	380	80	14	42.2	39.5	1971.9	55.3	55.6
	11 am	520	100	16	46.4	42.1	1968.8	53.7	55.2
	12 noon	580	120	18	45.5	41.8	1966.0	48.0	47.4
	1 pm	620	140	19	48.0	44.3	1963.6	45.8	46.9
	2 pm	580	120	22	54.3	47.3	1961.8	45.0	46.5
	3 pm	340	80	21	51.7	42.3	1960.1	42.2	43.0
	4 pm	200	40	20	43.5	38.7	1959.3	44.4	43.5
	5 pm	60	20	19	36.4	35.3	1959.1	43.2	42.5

(Continued)

TABLE 5.1 (Continued)

Experimental Observation of a Greenhouse Dryer for Jaggery Drying Under the Passive Mode

Day Drying	Time	$I(t)$ (W/m²)	$I(d)$ (W/m²)	T_a (°C)	T_j (°C)	T_e (°C)	$Mj \times 10^{-3}$ kg	γ_r (%)	γ_a (%)
3	10 am	420	60	14	42.2	39.5	1959.1	60.0	60.5
	11 am	600	100	17	46.4	42.1	1957.1	40.0	40.2
	12 noon	620	100	20	45.5	41.8	1954.8	33.5	31.3
	1 pm	680	100	21	48.0	44.3	1952.6	34.1	33.1
	2 pm	580	100	24	54.3	47.3	1950.6	32.3	32.7
	3 pm	440	60	24	51.7	42.3	1949.1	29.8	29.5
	4 pm	260	40	22	43.5	38.7	1948.4	33.1	32.6
	5 pm	60	20	21	36.4	35.3	1947.4	33.8	33.6
4	10 am	340	80	12	43.7	41.5	1947.4	55.8	56.4
	11 am	500	100	19	45.9	43.9	1946.2	42.0	41.2
	12 noon	640	100	20	51.9	47.7	1944.3	36.6	36.2
	1 pm	620	100	22	49.9	48.1	1942.5	35.3	35.2
	2 pm	560	100	24	51.9	47.1	1940.8	30.8	29.5
	3 pm	300	80	24	48.9	43.9	1939.7	36.9	36.7
	4 pm	200	60	22	44.3	40.1	1938.9	42.3	42.5
	5 pm	60	20	20	40.2	37.3	1938.5	44.8	45.0

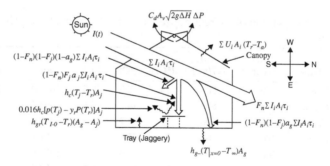

FIGURE 5.2
Thermal flow of the greenhouse dryer.

The data predicted by the thermal model was valided by experimental data. The study shows that both agree with high level of accuracy.

Ahmad and Prakash (2019) studied variations of convective heat transfer coefficients in the thermal storage based greenhouse dryer. A sensible thermal storage concept is being applied to the floor of the greenhouse dryer under passive mode. Four different floor conditions were selected, namely earth floor, concrete floor, gravel floor, and black gravel floor. The study reveals that the dryer operating over a black-painted gravel floor shows the highest value of convective heat transfer coefficient as compared to the other floor conditions.

5.2.2 Convective Mass Transfer

Convective mass transfer takes place from the external surface of the product to the hot air inside. During this transfer, the moisture in the product is reduced through the process of evaporation. The convective mass transfer coefficient (W/m^2K) is one of the important parameters of thermal modelling.

Kumar and Tiwari (2006c) studied the effect of the convective mass transfer coefficient over changes in the shape and size of the jaggery in the solar dryer. Three different sizes of jaggery were taken into consideration for the study. The sizes were $0.03 \times 0.03 \times 0.01$ m^3, $0.03 \times 0.03 \times 0.02$ m^3, and $0.03 \times 0.03 \times 0.03$ m^3. For the study, two sets of different weights of jaggery were selected, namely 0.75 kg and 2.0 kg. The dryer was operated in both modes of heat transfer, namely the natural convection and forced convection of heat transfer, as shown in Figure 5.3.

(a)

(b)

FIGURE 5.3
Photograph of the solar greenhouse dryer under (a) passive mode and (b) active mode.

Twelve sets of experiments were performed in both types of dryer. The experimental readings along with a computation for $0.03 \times 0.03 \times 0.01$ m^3 for 2.0 kg is shown in Table 5.2.

The study revealed that for a $0.03 \times 0.03 \times 0.02$ m^3 sample, the passive mode dryer showed a superior value of the convective mass transfer coefficient compared to the active mode of the greenhouse dryer. However, in the case of the $0.03 \times 0.03 \times 0.03$ m^3 sample, the active mode of greenhouse dryer showed a superior value compared to the passive mode of dryer.

TABLE 5.2

(a)–(b) Experimental Readings Along with Computation for $0.03 \times 0.03 \times 0.01$ m^3 for 2.0 kg for Both Passive and Active Modes of the Dryer

Day drying	Time	T_j (°C)	T_e (°C)	$\dot{m}_{ev} \times 10^{-3}$ kg	γ(%)	Gr $\times 10^5$	Pr	C	n	Nu	h_c (W/m^2 °C)
(a) Passive mode											
1	10–11 am	45.9	43.9	1.2	42.0	4.0	0.7	1.01	0.13	5.01	1.98
	11 am-12	51.9	47.7	1.9	36.6	6.24	0.7			5.3	2.1
	12–1 pm	49.9	48.1	1.8	35.3	4.72	0.7			5.12	2.02
	1–2 pm	51.9	47.1	1.7	30.8	6.86	0.7			5.37	2.13
	2–3 pm	48.9	43.9	1.1	36.9	6.25	0.7			5.3	2.08
	3–4 pm	44.3	40.1	0.8	42.3	5.06	0.7			5.16	2.0
	4–5 pm	40.2	37.3	0.4	44.8	3.86	0.7			4.99	1.91
2	10–11 am	52.2	39.3	2.0	40.0	11.02	0.7	0.99	0.12	5.24	2.06
	11 am-12	55.0	45.6	2.3	33.5	9.71	0.7			5.16	2.05
	12–1 pm	49.4	43.8	2.2	34.1	6.78	0.7			4.93	1.94
	1–2 pm	51.3	49.3	2.0	32.3	5.2	0.7			4.75	1.89
	2–3 pm	49.3	45.2	1.5	29.8	6.15	0.7			4.88	1.92
	3–4 pm	50.2	41.9	0.7	33.1	8.42	0.7			5.07	1.99
	4–5 pm	46.4	38.7	1.0	33.8	7.59	0.7			5.0	1.94
3	10–11 am	46.4	42.1	3.1	55.7	4.86	0.7	1.04	0.1	3.58	1.39
	11 am-12	45.5	41.8	2.8	48.0	4.66	0.7			3.57	1.39
	12–1 pm	48.0	44.3	2.4	45.8	5.01	0.7			3.59	1.40
	1–2 pm	54.3	47.3	1.8	45.0	7.7	0.7			3.57	1.49
	2–3 pm	51.7	42.3	1.7	42.2	8.86	0.7			3.8	1.49
	3–4 pm	43.5	38.7	0.8	44.4	5.26	0.7			3.61	1.39
	4–5 pm	36.4	35.3	0.2	43.2	2.56	0.7			3.37	1.27
4	10–11 am	44.2	33.8	6.8	45.8	8.59	0.7	1.01	0.06	2.45	0.93
	11 am-12	54.7	43.5	6.2	37.7	10.48	0.7			2.49	0.97
	12–1 pm	55.6	45.9	4.6	35.6	9.87	0.7			2.48	0.97
	1–2 pm	58.6	49.3	3.9	34.0	10.26	0.7			2.48	0.98
	2–3 pm	54.9	46.2	3.0	33.1	9.33	0.7			2.47	0.97
	3–4 pm	50.6	43.3	1.9	34.2	7.87	0.7			2.44	0.95
	4–5 pm	45.5	40.3	1.7	38.9	5.89	0.7			2.39	0.92

(Continued)

TABLE 5.2 (Continued)

(a)–(b) Experimental Readings Along with Computation for $0.03 \times 0.03 \times 0.01$ m^3 for 2.0 kg for Both Passive and Active Modes of the Dryer

Day drying	Time	T_i (°C)	T_e (°C)	$\dot{m}_{ev} \times 10^{-3}$ kg	$\gamma(\%)$	Gr × 10^5	Pr	C	n	Nu	h_c (W/m^2 °C)
(b) Active Mode											
1	10–11 am	38.9	37.1	1.6	47.6	2.24	0.7	0.98	0.16	3.38	1.23
	11 am-12	48.3	41.7	3.2	42.6	2.16	0.7			3.36	1.25
	12–1 pm	44.3	42.4	2.9	37.5	2.18	0.7			3.36	1.25
	1–2 pm	44.4	42.5	1.7	34.5	2.18	0.7			3.36	1.25
	2–3 pm	42.4	40.2	1.4	32.1	2.20	0.7			3.37	1.24
	3–4 pm	38.8	34.8	0.8	33.0	2.26	0.7			3.38	1.23
	4–5 pm	35.2	30.3	0.2	39.4	2.31	0.7			3.39	1.22
2	10–11 am	39.2	37.8	1.8	61.4	2.24	0.7	0.99	0.13	2.57	0.94
	11 am-12	46	44.1	2.5	52.4	2.16	0.7			2.56	0.95
	12–1 pm	47.4	43.3	2.3	53.6	2.15	0.7			2.56	0.96
	1–2 pm	49.3	45.1	2.6	49.4	2.13	0.7			2.56	0.96
	2–3 pm	46.1	42.2	1.7	47.6	2.17	0.7			2.56	0.95
	3–4 pm	43.2	38.3	1.1	53.7	2.21	0.7			2.57	0.95
	4–5 pm	38.9	36.6	0.5	54.7	2.25	0.7			2.58	0.94
3	10–11 am	34.2	31.1	4.8	49.5	2.31	0.7	1.0	0.12	2.59	0.93
	11 am-12	36.5	33.7	4.3	44.4	2.28	0.7			2.58	0.93
	12–1 pm	46.8	39.4	4.1	41.6	2.18	0.7			2.57	0.95
	1–2 pm	57.7	42.2	3.9	44.2	2.1	0.7			2.56	0.96
	2–3 pm	52.2	41.2	2.2	41.7	2.14	0.7			2.56	0.96
	3–4 pm	43.9	37.6	1.7	43	2.21	0.7			2.57	0.95
	4–5 pm	40.1	36.2	0.6	55.5	2.24	0.7			2.58	0.94
4	10–11 am	40.7	42.2	2.6	52	2.20	0.7	0.97	0.1	2.06	0.76
	11 am-12	44.7	44.4	3.4	46.2	2.16	0.7			2.06	0.77
	12–1 pm	48.7	46.6	4.2	41.3	2.13	0.7			2.06	0.77
	1–2 pm	46.1	46.0	4.8	43.0	2.15	0.7			2.06	0.77
	2–3 pm	44.1	42.3	4.0	36.8	2.18	0.7			2.06	0.76
	3–4 pm	41.1	36.4	1.9	34.8	2.24	0.7			2.07	0.76
	4–5 pm	36.5	33.0	0.1	37.2	2.29	0.7			2.07	0.75

Kumar and Tiwari (2007) evaluated convective mass transfer coefficients for different weights of onion flakes. The experiment was conducted using three modes of drying, namely the open sun dryer, and the active and passive modes of greenhouse dryer. Three sets of onions of weights 0.3 kg, 0.6 kg, and 0.9 kg were selected to dry in all three modes of drying. The experimental data along with the computation parameters for the 0.9 kg sample is presented in Table 5.3

The study reveals that the convective mass transfer coefficient is directly proportional to the mass and thickness of the onion flakes. Hence 0.9 kg of onion flakes show the highest value of convective mass transfer coefficient compared to the other weights.

TABLE 5.3

(a)–(c) Experimental Data Along with Computation Parameters for the 0.9 Kg of The Onion Sample Where (A) Open Sun Drying, (B) Passive Mode Drying, and (C) Active Mode Drying were Used

Time	T_c (°C)	T_e (°C)	$\dot{m}_{ev} \times 10^{-3}$ kg	γ (%)	Gr $\times 10^5$	Pr	C	n	Nu	h_c (W/m² °C)
(a) Open sun drying
8–9 am	26.0	30.1	9.8	40.8	3.28	0.69	0.899	0.164	6.86	2.42
9–10 am	33.3	37.2	18.8	32.1	3.69	0.69			6.99	2.52
10–11 am	35.0	40.1	45.2	31.5	4.28	0.69			7.16	2.60
11 am-12	35.9	42.4	49.1	26.6	5.06	0.69			7.36	2.69
12–1 pm	34.8	43.3	54.9	24.3	5.86	0.69			7.54	2.75
1–2 pm	32.9	41.7	47.9	22.8	6.0	0.69			7.57	2.75
2–3 pm	34.7	40.3	39.8	22.4	4.83	0.69			7.31	2.65
3–4 pm	33.5	35.3	30.3	23.1	3.12	0.69			6.8	2.45
4–5 pm	28.0	30.5	13.5	37.0	2.68	0.69			6.64	2.35
5–6 pm	23.9	26.6	3.0	50.7	2.33	0.69			6.49	2.27
6–7 pm	22.2	25.3	1.0	55.2	2.42	0.69			6.53	2.28
7–8 pm	20.7	23.9	0.1	57.8	2.42	0.69			6.53	2.27
8–9 pm	20.5	23.8	0.3	60.0	2.43	0.69			6.53	2.27
9–10pm	19.9	23.5	0.9	59.7	2.6	0.69			6.6	2.29
10–11 pm	19.0	22.8	0.9	60.4	2.38	0.69			6.51	2.26
11pm-12	19.0	22	1.1	61.6	2.25	0.69			6.45	2.23
12–1 am	18.9	21.8	1.3	62.3	2.13	0.69			6.39	2.21
1–2 am	18.8	21.7	1.0	63.4	2.12	0.69			6.38	2.21
2–3 am	18.5	21.9	1.1	64.8	2.25	0.69			6.45	2.23
3–4 am	18.3	21.4	1.2	78.2	1.91	0.69			6.28	2.17
4–5 am	18.4	21.3	1.1	82.0	1.83	0.69			6.28	2.17
5–6 am	19.5	21.3	1.0	83.0	1.53	0.69			6.24	2.15
6–7 am	22.1	21.9	1.7	83.8	1.48	0.69			6.05	2.09
7–8 am	26.2	24.5	0.6	84.2	2.72	0.69			6.02	2.1
8–9 am	30.8	30.0	0.2	57.0	3.88	0.69			7.05	2.35
9–10 am	36.8	40.9	9.8	34.3	5.13	0.69			7.38	2.57
10–11 am	40.1	45.1	26.7	19.6	4.65	0.69			7.26	2.72
11 am-12	39.8	43.9	36.7	22.2	4.75	0.69			7.29	2.67
12–1 pm	38.7	43.3	28.2	21.8	4.85	0.69			7.31	2.67
1–2 pm	34.1	40.0	32.2	24.0	3.32	0.69			6.87	2.65
2–3 pm	37.1	38.4	24.4	21.3	2.55	0.69			6.58	2.5
3–4 pm	33.6	34.3	20.9	26.0	2.32	0.69			6.48	2.37
4–5 pm	28.7	30.6	7.5	41.2	2.12	0.69			6.42	2.3
(b) Passive mode dryer
8–9 am	27.5	29.5	4.7	42.0	2.3	0.69	1.067	0.140	5.76	2.04
9–10 am	32.2	37.1	13.2	33.3	4.01	0.69			6.23	2.24
10–11 am	36.9	39.6	36.0	33.1	3.41	0.69			6.09	2.22
11 am-12	40.8	42.0	40.7	28.1	3.42	0.69			6.09	2.24

(Continued)

TABLE 5.3 (Continued)

(a)–(c) Experimental Data Along with Computation Parameters for the 0.9 Kg of The Onion Sample Where (A) Open Sun Drying, (B) Passive Mode Drying, and (C) Active Mode Drying were Used

Time	T_c (°C)	T_e (°C)	$\dot{m}_{ev} \times 10^{-3}$ kg	γ (%)	Gr × 10⁵	Pr	C	n	Nu	h_c (W/m² °C)
12–1 pm	42.1	43.6	46.9	26.2	3.77	0.69			6.18	2.28
1–2 pm	42.5	42.8	42.1	25.0	3.45	0.69			6.1	2.25
2–3 pm	37.5	39.3	32.3	23.7	3.48	0.69			6.11	2.22
3–4 pm	35.4	39.6	26.6	25.2	4.2	0.69			6.27	2.28
4–5 pm	30.2	32.7	14.9	40.1	2.69	0.69			5.89	2.1
5–6 pm	26.0	29.1	5.4	53.8	2.48	0.69			5.88	2.06
6–7 pm	24.2	27.2	3.6	56.4	2.35	0.69			5.78	2.03
7–8 pm	22.8	26.1	2.6	60	2.42	0.69			5.8	2.03
8–9 pm	22.1	25.2	1.8	61.6	2.29	0.69			5.76	2.01
9–10 pm	21.8	25.2	1.2	60.7	2.36	0.69			5.78	2.01
10–11 pm	21.2	25.0	2.0	62.1	2.33	0.69			5.78	2.02
11pm-12	20.6	24.4	0.6	63.7	2.41	0.69			5.8	2.0
12–1 am	20.4	24.0	1.0	65	2.28	0.69			5.76	2.0
1–2 am	20.0	23.8	0.9	66.2	2.26	0.69			5.75	2.0
2–3 am	19.8	23.6	1.1	67.1	2.3	0.69			5.76	1.96
3–4 am	20.2	23.3	1.1	79.1	2.02	0.69			5.66	1.95
4–5 am	21.5	23	1.0	82.4	1.91	0.69			5.62	1.85
5–6 am	23.7	22.7	1.3	83.1	1.58	0.69			5.47	2.04
6–7 am	27.0	23.5	0.9	83.8	1.3	0.69			5.32	2.23
7–8 am	36.3	25.7	1.5	84.5	1.26	0.69			5.3	2.33
8–9 am	39.5	30.0	10.4	58.8	2.31	0.69			5.77	2.29
9–10 am	42.4	39.9	26.5	36.2	3.58	0.69			6.13	2.26
10–11 am	43.4	43.2	37.1	21	4.53	0.69			6.34	2.25
11 am-12	40.1	44.1	28.0	24.5	3.96	0.69			6.22	2.23
12–1 pm	38.5	43.2	31.9	24.6	3.59	0.69			6.13	2.21
1–2 pm	35.0	41.6	22.1	25.6	3.56	0.69			6.13	2.15
2–3 pm	30.8	40.1	19.1	23.2	3.52	0.69			6.12	2.10
3–4 pm	35.0	37.8	15.5	28.2	3.49	0.69			6.11	2.08
4–5 pm	30.8	32.8	9.5	45.3	2.35	0.69			5.78	2.06
(c) Active mode dryer										
8–9 am	16.1	17.9	5.4	53.6	2.54	0.69	1.004	0.271	7.66	2.62
9–10 am	29.2	28.5	15.6	35.5	2.37	0.69			7.51	2.66
10–11 am	38.7	38.6	28.8	32.0	2.24	0.69			7.39	2.69
11 am-12	33.4	34.8	40.8	26.9	2.3	0.69			7.44	2.68
12–1 pm	41.0	43.0	40.8	26.5	2.19	0.69			7.35	2.7
1–2 pm	34.3	35.5	44.2	24.5	2.28	0.69			7.44	2.68
2–3 pm	41.0	29.3	37.0	23.6	2.37	0.69			7.51	2.7

(Continued)

TABLE 5.3 (Continued)

(a)–(c) Experimental Data Along with Computation Parameters for the 0.9 Kg of The Onion Sample Where (A) Open Sun Drying, (B) Passive Mode Drying, and (C) Active Mode Drying were Used

Time	T_c (°C)	T_e (°C)	$\dot{m}_{ev} \times 10^{-3}$ kg	γ (%)	Gr × 10⁵	Pr	C	n	Nu	h_c (W/m² °C)
3–4 pm	34.3	26.5	30.8	27.6	2.41	0.69			7.55	2.68
4–5 pm	28.4	21.1	17.4	45.2	2.5	0.69			7.63	2.66
5–6 pm	24.8	27.8	8.1	65.6	2.4	0.69			7.54	2.65
6–7 pm	17.8	27.1	4.7	68.1	2.41	0.69			7.56	2.63
7–8 pm	24.6	26.5	4.3	66.5	2.42	0.69			7.57	2.65
8–9 pm	24.1	25.6	3.1	74.6	2.43	0.69			7.57	2.65
9–10pm	23.5	25.3	2.2	74	2.44	0.69			7.58	2.65
10–11 pm	22.8	25.0	2.1	75.3	2.44	0.69			7.58	2.64
11pm-12	22.4	24.8	1.9	76.2	2.44	0.69			7.58	2.64
12–1 am	22.1	24.3	1.6	77	2.45	0.69			7.59	2.64
1–2 am	22.1	23.7	1.4	77.6	2.46	0.69			7.6	2.64
2–3 am	21.6	23	1.2	78	2.47	0.69			7.6	2.64
3–4 am	21.1	23	1.2	78.4	2.47	0.69			7.6	2.64
4–5 am	20.2	22.7	1.2	78.8	2.47	0.69			7.6	2.64
5–6 am	20.4	22.7	0.4	58.2	2.47	0.69			7.58	2.64
6–7 am	20.3	22.7	0.7	34.4	2.45	0.69			7.55	2.64
7–8 am	22.2	24.0	0.7	30.2	2.41	0.69			7.42	2.64
8–9 am	25.6	25.5	2.3	27.0	2.27	0.69			7.35	2.64
9–10 am	37.5	34.8	11.6	23.6	2.19	0.69			7.33	2.65
10–11 am	43.6	41.1	28.4	26.0	2.17	0.69			7.45	2.69
11 am-12	43.7	44.2	34.8	27.8	2.30	0.69			7.51	2.71
12–1 pm	33.6	33.6	35.9	31.8	2.37	0.69			7.41	2.71
1–2 pm	28.1	29.5	29.4	36.2	2.26	0.69			7.43	2.68
2–3 pm	36.9	36.8	20.2	40.2	2.28	0.69			7.45	2.66
3–4 pm	35.0	35.5	18.6	43.2	2.31	0.69			7.46	2.69
4-5 pm	30.0	32.2	10.1	51.0	2.34	0.69			7.48	2.68

5.3 Modeling Procedure for Convective Heat Transfer (CHT)

In this section, a systematic procedure is developed to model the convection developed by Singh and Tiwari (2020).

The coefficient of CHTC represented by h_c is expressed as follows:

$$h_c = \frac{K_c}{X_l} \mathrm{Nu} \tag{5.1}$$

Nu = Nusselt number
K_c = thermal conductivity of the humid air (W/m K)
X_l = characteristic length (m)

The rate of heat is required to remove moisture from the product. It is expressed mathematically as follows:

$$\dot{H}_e = 0.016 h_c \left[P(T_a) - \gamma P(T_b) \right] \tag{5.2}$$

$P(T_i)$ = partial vapor pressure,

$$P(T_i) = \exp\left[25.317 - \frac{5144}{T_i + 273.15} \right] \tag{5.3}$$

$P(T_a)$ = partial vapor pressures of air at temperatures T_a.
$P(T_b)$ = partial vapor pressures of air at environment temperature T_b

Equation (5.2) can also be expressed as follows:

$$\dot{H}_e = h_e (T_a - T_b) \tag{5.4}$$

h_e = evaporative heat transfer coefficient (Qiu et al., 2020)

$$h_e = \frac{16.273 \times 10^{-3} h_c \left(P(T_a) - \gamma P(T_b) \right)}{T_a - T_b} = \frac{0.016 h_c \left(P(T_a) - \gamma P(T_b) \right)}{T_a - T_b} \tag{5.5}$$

From Equations (5.1) and (5.2):

$$\dot{H}_e = 0.016 \frac{K_V}{X_l} \text{Nu} \left[P(T_a) - \gamma P(T_b) \right] \tag{5.6}$$

The moisture evaporation (m_v) of the crop can be evaluated by the following expression:

$$m_v = \frac{\dot{H}_e}{\lambda} A_t t = 0.016 \frac{K_V}{X_l \lambda} \text{Nu} \left[P(T_a) - \gamma P(T_b) \right] A_t t \tag{5.7}$$

Where m_v is in kg, λ is latent heat of vaporization (J/kg), A_t is the area of tray (m²), and t is time (sec):
After rearranging, Equation (5.7) can be written as follows:

$$\frac{m_v}{z} = \text{Nu} \tag{5.8}$$

Where z is equal to

$$z = 0.016 \frac{K_V}{X_l \lambda} \left[P(T_a) \quad \gamma P(T_b) \right] A_t t \tag{5.9}$$

5.3.1 Case I: Natural Convection

In the natural convection mode of heat transfer, the Nusselt number (Nu) is the function of the Grashof number (Gr) and Prandtl number (Pr) are calculated for the solar dryer. Mathematically, it is expressed as follows:

$$\mathrm{Nu} = \frac{h_c X_l}{K_v} = C(\mathrm{Gr.Pr})^n \tag{5.10}$$

C and n are the constants.

From Equations (5.9) and (5.10):

$$\frac{m_v}{Z} = C(\mathrm{GrPr})^n \tag{5.11}$$

Taking the logarithm in Equation (5.11):

$$\ln\left[\frac{m_{ev}}{Z}\right] = n\ln(\mathrm{GrPr}) + \ln C \tag{5.12}$$

This is the form of a linear equation:

$$Y = mX_0 + C_0$$

$$\text{where } Y = \ln\left[\frac{m_{ev}}{Z}\right], m = n, X_0 = \ln[\mathrm{GrPr}] \text{ and } C_0 = \ln C \tag{5.13}$$

$$C = e^{C_0} \tag{5.14}$$

The linear regression analysis is done with the constants evaluation:

$$m = n = \frac{N\sum X_0 Y - \sum X_0 \sum Y}{N\sum X_0^2 - (\sum X_0)^2} \quad \text{and} \quad C_0 = \frac{\sum X_0^2 \sum Y - \sum X_0 \sum X_0 Y}{N\sum X_0^2 - (\sum X_0)^2} \tag{5.15}$$

5.3.2 Case II: Forced Convection

In the forced convection mode of heat transfer, the Nusselt number (Nu) is the function of the Reynolds number (Re) and Prandtl number (Pr). Mathematically, it is expressed as follows:

$$\mathrm{Nu} = \frac{h_c X_l}{K_V} = C'\left(\mathrm{RePr}\right)^{n'} \tag{5.16}$$

From Equations (5.11) and (5.16):

$$\frac{m_v}{Z} = C'\left(\mathrm{RePr}\right)^{n'} \tag{5.17}$$

Equation (5.17) is equivalent to Equation (5.11).

h_c can be evaluated by Equation (5.16) for forced convection.

The C' and n' constants are calculated in a similar way to the natural convection mode.

5.4 Thermal Modeling of Various Types of Dryers

Thermal modelling is an important mathematical/theoretical modeling technique for solar drying. It is very useful to model a complex process of drying where heat and mass transfer take place simultaneously. Thermal modeling is being applied with all types of solar dryer.

5.4.1 Thermal Modeling of Open Sun Drying

In the process of thermal modelling of open sun drying, the following assumptions are required to write the energy balance equation:

- The heat capacity of the drying tray and surrounding air is to be ignored.
- Conduction heat transfer is to be ignored.
- The flow of heat is unidirectional.
- There is no stratification in the depth of the crop layer.

The energy balance equation is written based on the first law of thermodynamics.

5.4.1.1 Crop Surface

$$\alpha_c I(t) A_t - h_{rc}\left(T_a - T_b\right) A_t - 0.016 h_c \left[P(T_a) - \gamma P(T_b)\right] A_t$$
$$-h_i\left(T_a - T_{ab}\right) A_t = M_c C_c \frac{dT_c}{dt} \tag{5.18}$$

T_{ab} = ambient air temperature (°C)
T_a = crop temperature (°C)
h_l = overall bottom heat loss coefficient (W/m² K)
T_b = temperature just above the crop surface (°C)
$I(t)$ = solar radiation incident (W/m²)
h_{rc} = radiative heat transfer coefficient (RHTC) and conductive heat transfer
         coefficient (CHTC) from the crop surface to its surroundings (W/m² K)
α_c = absorptivity of the crop surface
C_c = specific heat of the crop (J/kgK)

5.4.1.2 Moist Air above the Crop

$$h_{rc}(T_a - T_b)A_t + 0.016h_c\left[P(T_a) - \gamma P(T_b)\right]A_t = h_2(T_b - T_{ab})A_t \qquad (5.19)$$

Removed moisture, or evaporation (MeV), could be assessed as

$$h_2 = 5.7 + 3.8\,V \text{ is CHTC and RHTC}$$

V is wind velocity (m/s)
The moisture evaporated in open sun drying or natural solar drying is calculated as follows:

$$m_v = 0.016\frac{h_e}{\lambda}\left[P(T_a) - \gamma P(T_b)\right]A_t t \qquad (5.20)$$

The operating temperature ranges between 32°C and 50°C for open sun drying where partial pressure is evaluated.

$$P(T) = R_a T + R_b \qquad (5.21)$$

where R_a and R_b are constant, which is taken from the steam table.
With the help of Equations (5.18) and (5.19), the new equation becomes as follows:

$$\alpha_c I(t)A_t - h_{rc}(T_a - T_b)A_t - 0.016h_c\left[(R_a T_a + R_2) - \gamma(R_b T_b + R_2)\right]$$
$$A_t - h_i(T_a - T_b)A_t = M_c C_c \frac{dT_c}{dt} \qquad (5.22)$$

$$h_c\left(T_a - T_b\right) + 0.016h_c\left(R_aT_a + R_b\right) - 0.016h_c\gamma\left(R_aT_b + R_b\right) = h_2\left(T_b - T_{ab}\right)$$
$$h_c\left(T_a - T_b\right) + 0.016h_c\left(R_aT_a + R_b\right) - 0.016h_c\gamma\left(R_aT_b + R_b\right) = h_2\left(T_b - T_{ab}\right) \quad (5.23)$$

Equation (5.23) can be rewritten as

$$T_b = \frac{\left(h_{rc} + 0.016h_cR_a\right)T_a + R_b\left[0.016h_c\left(1-\gamma\right)\right] + h_2T_{ab}}{h_{rc} + 0.016h_c\gamma + h_2} \quad (5.24)$$

And Equation (5.22) can be rewritten as

$$\alpha_c I(t)A_t - h_iA_tT_{ab} - \left[\left(h_{re} + h_i\right)A_t + 0.016h_cR_aA_t\right]T_a + \left(h_{rc} + 0.016h_cR_a\gamma\right)A_tT_b$$
$$-0.016h_cR_bA_t\left(1-\gamma\right) = M_cC_c\frac{dT_a}{dt} \quad (5.25)$$

Substituting Equations (5.24) in (5.25), we get

$$\left\{\left[\alpha_c I(t)A_t + h_iA_tT_{ab}\right] - \left[\left(h_{rc} + h_i\right)A_t + 0.016h_cR_aA_t\right]\right\}T_a$$
$$+ \left[\left(h_{rc} + 0.016h_cR_a\gamma\right)A_t\right]\left\{\frac{\left(h_{rc} + 0.016h_cR_a\right)T_a + R_b\left[0.016h_c\left(1-\gamma\right)\right] + h_2T_{ab}}{h_{rc} + 0.016h_c\gamma R_a + h_2}\right\}$$
$$-0.016h_cR_bA_t\left(1-\gamma\right) = M_cC_c\frac{dT_a}{dt} \quad (5.26)$$

$$\frac{dT_a}{dt} + \left\{\frac{\left[\left(h_{rc} + h_i\right)A_t + 0.016h_cR_aA_t\right] - \dfrac{A_t\left(h_{rc} + 0.016h_cR_a\gamma\right)\left(h_{rc} + 0.016h_cR_a\right)}{h_{rc} + 0.016h_cR_a}}{M_cC_c}\right\}T_a$$
$$= \frac{1}{M_cC_c}\left[\alpha_c I(t)A_t + h_iA_tT_{ab}) + \frac{\left(h_{rc} + 0.016h_cR_a\gamma\right)\cdot A_t\left[R_b\left(0.016h_c\left(1-\gamma\right)\right) + h_2T_{ab}\right]}{h_{rc} + 0.016h_c\gamma R_a + h_2}\right.$$
$$\left. -(0.016h_cR_bA_t\left(1-\gamma\right)\right] \quad (5.27)$$

This is the form of the first-order differential equation

$$\frac{dT_a}{dt} + aT_a = f(t) \tag{5.28}$$

Where

$$a = \frac{\left[(h_{rc} + h_i)A_t + 0.016h_cR_aA_t\right] - \dfrac{A_t(h_{rc} + 0.016h_cR_a\gamma)(h_{rc} + 0.016h_cR_a)}{h_{rc} + 0.016h_c\gamma R_a + h_2}}{M_cC_c}$$

And

$$f(t) = \frac{1}{M_cC_c}\Big[\alpha_cI(t)A_t + h_iA_tT_{ab}$$

$$+ \frac{(h_{rc} + 0.016h_cR_a\gamma)\cdot A_t\left[R_b(0.016h_c(1-\gamma)) + h_2T_{ab}\right]}{h_{rc} + 0.016h_c\gamma R_a + h_2} - 0.016h_cR_bA_t(1-\gamma)\Big]$$

The analytical solution of Equation (5.28) is

$$T_a = \frac{\overline{f(t)}}{a}\left(1 - e^{-a\Delta t}\right) + T_{a0}\cdot e^{-a\Delta t} \tag{5.29}$$

And from Equation (5.21) we get

$$m_{ev} = 0.016\frac{h_c}{\lambda}\left[(R_aT_a + R_b) - \gamma(R_aT_b + R_b)\right]A_t t \tag{5.30}$$

It predicted and detected that there is an agreement between the theoretical and the evaluated results.

5.4.1.3 Analysis for Steady State Condition

In the steady state condition, $\frac{dT_c}{dt} = 0$ is to be taken from Equations (5.22) and (5.23), the new equation is being developed as

$$\alpha_cI(t)A_t - h_{rc}(T_a - T_b)A_t - 0.016h_c\left[(R_aT_a + R_b) - \gamma(R_aT_b + R_b)\right]A_t$$
$$- h_i(T_a - T_{ab})A_t = 0 \tag{5.31}$$

$$(h_{rc} + 0.016h_cR_a)T_a + R_b0.016h_c(1-\gamma) + h_2T_{ab} = (h_{rc} + 0.016h_cR_a\gamma + h_2)T_b \tag{5.32}$$

From Equation (5.31), the expression of T_a can be written as

$$T_a = \frac{\alpha_c I(t) + h_i T_{ab} + (h_{rc} + 0.016 h_c R_a \gamma) T_b - 0.016 h_c R_b (1 - \gamma)}{h_{rc} + 0.016 h_c R_a + h_i} \qquad (5.33)$$

Substituting Equations (5.33) in (5.32) we can get the expression of T_e as

$$T_b = \frac{H \alpha_c}{h_T} I(t) + \frac{H h_i + h_2}{h_T} T_a + \frac{H_i}{h_T} 0.016 h_c (1 - \gamma) R_a \qquad (5.34)$$

Where

$$h_T = \frac{(h_{rc} + 0.016 h_c R_a)(h_{rc} + 0.016 h_c R_a \gamma)}{h_{rc} + 0.016 h_c R_a + h_i} + h_{rc} + 0.016 h_c R_a \gamma + h_2$$

$$H = \frac{h_{rc} + 0.016 h_c R_a}{h_{rc} + 0.016 h_c R_a + h_i} \quad \text{and} \quad H_i = \frac{h_i}{h_{rc} + 0.016 h_c R_a + h_i}$$

T_a from the expression mentioned above (Equation 5.34) is also equivalent to the solar temperature for open sun drying.

5.4.2 Thermal Modeling of Greenhouse Drying Systems (TMGDS)

Thermal modeling is being used to predict the crop temperature (T_a), room temperature (T_r), and moisture evaporation (m_v) during the process of drying in the greenhouse dryer under both active and passive modes. For the writing of an energy balance equation of the various components based on the first law of thermodynamics, the following assumptions are to be made:

- The heat capacity of the transparent and non-transparent walls, frame, and structural materials is to be ignored.
- Stratification, heat absorption capacity, and heat capacity of the room air temperature is to be ignored.
- This methodology is applicable to thin-layer drying.
- Shrinkage of the crop is to be ignored.

5.4.2.1 Greenhouse Dryers under Natural Convection Mode

The energy balance equation of the various important parts of greenhouse dryer in passive mode is presented below.

(a) Crop Surface

$$(1-F_n)(F_c)(\alpha_c)\Sigma I_i A_i \tau_i$$
$$= M_c C_c \frac{dTc}{dt} h(T_a - T_r)A_c + 0.016 h_c \left[P(T_a) - \gamma_r P(T_r)\right]A_c \qquad (5.35)$$

F_c = fraction of solar radiation on crop surface
F_n = fraction of solar radiation on the north wall
T_r = greenhouse room air temperature (°C)
T_a = crop temperature (°C)

(b) Ground Surface

$$(1-F_n)(1-F_c)(\alpha_g)\Sigma I_i A_i \tau_i$$
$$= h_{g\infty}\left(T\big|_{y=0} - T_\infty\right)A_g + h_{gr}(T\big|_{y=0} - T_r)\left(A_g - A_c\right) \qquad (5.36)$$

T_∞ = underground soil temperature (°C)
α_g = ground absorptivity
$h_{g\infty}$ = HTC floor to underground (W/m² K)
h_{gr} = HTC between greenhouse floor and greenhouse room (W/m² K)

(c) Greenhouse Room
T_r = temperature difference (°C)
T_{ab} = ambient air temperature (°C)
C_d = coefficient of diffusion

The energy balance is represented as:

$$(1-F_n)(1-F_c)(1-\alpha_g)\Sigma I_i A\tau_i + h(T_a - T_r)A_t + 0.016 h_c \left[P(T_a) - \gamma_r P(T_r)\right]$$
$$A_c + h_{gr}\left(T\big|_{y=0} - T_r\right)\left(A_g - A_c\right)$$
$$= C_d A_V \sqrt{2g\Delta H} \times \Delta P + \Sigma U_i A_i \left(T_r - T_{ab}\right)$$

$$(5.37)$$

$C_d A_V \sqrt{2g\Delta H} \times \Delta P$ is a loss of heat through the outlet vent (Tiwari, 2003).

$$\Delta H = \frac{\Delta P}{\rho_r g} \qquad (5.38)$$

$$\Delta P = P(T_r) - \gamma_a P(T_{ab}) \qquad (5.39)$$

5.4.2.2 Solution of the Thermal Model for Greenhouse Dryers under Passive Mode

Since all small-scale greenhouse dryers have an operating temperature which varies between 25°C to 55°C, the vapor pressure can be expressed as

$$P(T) = R_1T + R_2 \tag{5.40}$$

The convective heat transfer coefficient for the crop is being calculated based on the procedure described in Section 5.3 (Case I).

Equations (5.35), (5.36) and (5.39) can be simplified into a third-order polynomial equation with the help of Equations (5.36) and (5.37) to determine the greenhouse room temperature.

The third-order polynomial can be expressed as

$$AT_r^3 + BT_r^2 + CT_r + D = 0$$

Where

$$A = \left(\frac{2}{\rho_r}\right)(C_dA_v)^2 R_1^3$$

$$B = \left(\frac{2}{\rho_r}\right)(C_dA_v)^2 3R_1^2\left[R_2 - \gamma_a(R_1T_{ab} + R_2)\right]$$
$$- \left[hA_c + 0.016h_cA_r\gamma_rR_1 + (UA)_{g\circ} + \Sigma U_iA_i\right]$$

$$C = \left(\frac{2}{\rho_r}\right)(C_dA_v)^2 3R_1\left[R_2 - \gamma_a(R_1T_{ab} + R_2)\right]^2$$
$$+ 2\left[hA_c + 0.016h_cA_c\gamma_rR_1 + (UA)_{gm} + \Sigma U_1A_i\right]$$
$$\times \left[I_{effR} + I_{eff}H_c + T_c(hA_c + 0.016h_cA_cR_1) + T_{ab}(UA)_{g\infty} + T_a\Sigma U_1A_i + 0.016h_cA_cR_2(1-\gamma_r)\right]^{-2}$$

$$D = \left(\frac{2}{\rho_r}\right)(C_dA_v)^2\left[R_2 - \gamma_a(R_1T_{ab} + R_2)\right]^3$$
$$- \left[\begin{array}{l}I_{effR} + I_{effG}H_G + T_a(hA_c + 0.016h_cA_cR_1) + T_{ab}(UA)_{g\infty}\\ + T_{ab}\Sigma U_iA_i + 0.016h_cA_cR_2(1-\gamma_r)\end{array}\right]$$

With the help of the greenhouse air temperature (T_r) in Equation (5.35), the crop temperature T_a can be discovered by using the first-order differential equation, expressed as below

$$\frac{dT_a}{dt} + aT_a = f(t) \tag{5.41}$$

For the time 0 to t, the average $\overline{f(t)}$ from Equation (5.41) can be

$$T_{alt=0} = T_{co} \text{ is } T_a = \frac{\overline{f(t)}}{a}\left(1-e^{-at}\right)+T_{co}e^{-at} \qquad (5.42)$$

$$a = \frac{h_c A_c\left(1+0.016R_1\right)}{M_c C_c}$$

$$\overline{f(t)} = \frac{I_{eff}+h_c A\left[T-0.016\left\{R-\gamma\left(RT+R\right)\right\}\right]}{MC}$$

With the help of the above expression, the moisture evaporated can be calculated as

$$m_{vA} = 0.016\frac{h_c}{\lambda}\left[\left(R_1 T_a + R_2\right)-\gamma\left(R_1 T_r + R_2\right)\right]A_c t \qquad (5.43)$$

In the no-load condition, the coefficient of diffusion can be expressed as

$$\left(1-F_n\right)\left(1-\alpha_g\right)\Sigma I_i A_i \tau_i + h_{gr}\left(T\big|_{y=0} - T_r\right)A_g$$
$$= C_d A_v\sqrt{2g\Delta H}\times\Delta P + \Sigma U_i A_i\left(T_r - T_{ab}\right) \qquad (5.44)$$

5.4.2.3 Greenhouse Dryers under Forced Convection Mode

Energy balance equations are being developed for the various important components of greenhouse dryers under the forced convection mode.

In this condition, for both the crop surface and ground surface, the same equation is being applied here as was used for the greenhouse dryer in the natural convection mode.

(a) Greenhouse Room

The energy balance equation is being developed with the help of the number of air exchanges per hour (N) for the greenhouse dryer under active mode:

$$\left(1-F_n\right)\left(1-F_c\right)\left(1-\alpha_g\right)\Sigma I_i A_i \tau_i + h\left(T_a-T_r\right)A_c + 0.016h_c\left[P\left(T_c\right)-\gamma_r P\left(T_r\right)\right]A_c + h_{gr}\left(T\big|_{y=0}\right.$$
$$\left.-T_r\right)\left(A_g - A_c\right)$$
$$= 0.33NV\left(T_r - T_{ab}\right)+\Sigma U_i A_i\left(T_r - T_{ab}\right)$$

$$(5.45)$$

5.4.2.4 *Solution of the Thermal Model*

With the help of the greenhouse air temperature (T_r) in Equation (5.35), The crop temperature T_a can be found by using the first-order differential equation, expressed as below:

$$A_r = BT_a + CT_{ab} + D$$

$$A = \left[hA_c + 0.016h_cA\gamma_1R_1 + (UA)_{g-} + 0.33NV + \Sigma U_iA_i \right]$$

$$B = \left[hA_c + 0.016h_cAR_1 \right]$$

$$C = \left[(UA)_{se} + \Sigma U_iA_i + 0.33NV \right]$$

$$D = \left[I_{dFR} + I_{ffG}H_G + 0.016h_cA_c(1 - Y_r) \right]$$

All other parameters are to be evaluated based on the natural convection mode of greenhouse dryer.

These are the important derivatives for thermal modelling in the load condition:

$$I_{effc} = (1 - F_n)F_c\alpha_c \Sigma I_iA_i\tau_i$$

$$I_{effG} = (1 - F_n)(1 - F_c)\alpha_g \Sigma I_iA_i\tau_i$$

$$I_{effR} = (1 - F_n)(1 - F_c)(1 - \alpha_g) \Sigma I_iA_i\tau_i$$

$$II_G = \left[1 + \frac{h_{800}A_8}{h_{gr}(A_g - A_c)} \right]^{-1}$$

$$(UA)_{g\infty} = \left[\frac{1}{h_{gr}(A_g - A_c)} + \frac{1}{h_{g-h}A_g} \right]$$

$$h_r = \frac{\sigma\varepsilon \left[(T_c + 273.15)^4 - (T_r + 273.15)^4 \right]}{(T_a - T_r)}$$

$$\frac{1}{U} = \left(\frac{1}{h_1} \right) + \left(\frac{l_g}{k_g} \right) + \left(\frac{1}{h_2} \right)$$

$$\Sigma A_iU_i = U \Sigma A_i$$

$$h_1 = h_c + h_r + h_e$$

$$h = h_c + h_r$$

$$h_2 = 5.7 + 3.8V$$

$$h_e = \frac{16.273 \times 10^{-3}h_c \left(P(T_a) - \gamma_r P(T_r) \right)}{T_a - T_r}$$

$$P(T) = \exp\left(25.317 - \frac{5144}{273 + T} \right)$$

$$\overline{T}_\infty = 25°C \text{ (Khatry et al.,1978)}$$

5.4.3 Thermal Modeling of Indirect Solar Drying (ISD) Systems

The indirect solar dryer has two major components, namely the solar air heater and the drying chamber.

5.4.3.1 Solar Air Heater

The solar air heater is used to supply hot air in the drying chamber.

The energy balance equation for the different components of the solar air heater is expressed as follows:

(i) Absorber plate

The energy balance equation of the absorber plate is as follows:

$$\tau \alpha_p I W dx = U_t \left(T_p - T_{ab} \right) W dx + h_{pf} \left(T_p - T_f \right) W dx \tag{5.46}$$

Where

$$U_t = \left[\frac{1}{h_{pg}} + \frac{1}{h_{ga}} \right]^{-1}$$

τ = transmissivity of the glass cover
α_p = absorptivity of the absorber plate
T_p = absorber plate temperature (°C)
T_{ab} = ambient temperature (°C)
W = width of the absorber (m)
I = global solar intensity (W/m²)
U_t = overall top loss coefficient (W/m² K)
h_{pf} = convective heat transfer coefficient between plate and flowing inside air (W/m² K)
h_{pg} = convective heat transfer coefficient between plate and glass cover (W/m² K)
h_{ga} = convective heat transfer coefficient between ambient air and glass cover (W/m² K)

(ii) Air flow

The energy balance equation of the air flow inside the solar heater is as follows:

$$h_{pf} \left(T_p - T_f \right) W dx = \dot{m} C_f \frac{dT_f}{dx} dx + U_b \left(T_f - T_{ab} \right) W dx \tag{5.47}$$

Where U_b is the overall heat transfer coefficient of flowing air to ambient air (W/m² K).

5.4.3.2 Drying Chamber

The energy balance equation of the drying chamber is as follows:

$$\dot{m}C_f\left(T_{fo}-T_{fi}\right)=M_cC_c\frac{dT_c}{dt}+h\left(T_c-T_{ch}\right)A_{ch} \tag{5.48}$$

and

$$h\left(T_c-T_{ch}\right)A_{ch}=\dot{m}C_f\left(T_{ch}-T_{ab}\right)+h_sA_s\left(T_{ch}-T_{ab}\right) \tag{5.49}$$

5.5 Statistical Parameters

Statistics is the technique used to discover patterns in the large amounts of data. These are the prominent statistical parameters used in the thermal modeling of the dryer.

5.5.1 Root Mean Square Deviation/Root Mean Square Error

The root mean square deviation (RMSD) is also known as the root mean square error. It measures the difference between predicted and experimental values. It is expressed mathematically as:

$$RMSE/RMSD=\sqrt{\frac{\Sigma\left(p_{i_i}\right)^2}{N}} \tag{5.50}$$

Where

$$p_{i_i}=\left[\frac{X_{pr,i}-X_{ex,i}}{}\right]$$

$X_{ex,i}$ = *i*th experimental values
N = number of observations
$X_{pr,i}$ = predicted value
i = 1 to N

5.5.2 Coefficient of Correlation (*r*)

The coefficient of correlation develops the relationship between experimental and predicted values. It is denoted by 'r' and its value varies between -1 and 1.

$$r = \frac{\left(N\sum X_{ex,i}X_{pr,i}\right) - \left(\sum X_{ex,i}\right) \times \left(\sum X_{pr,i}\right)}{\left(\sqrt{N\sum X_{ex,i}^2 - \left(\sum X_{ex,i}\right)^2}\right)\left(\sqrt{N\sum X_{pr,i}^2 - \left(\sum X_{pr,i}\right)^2}\right)} \tag{5.51}$$

5.5.3 Coefficient of Determination

The coefficient of determination is denoted by R^2. It indicates how well the data fit the trendline curve. It is a tool that is used in validating the prediction model or hypotheses based on other related information.

$$R^2 = 1 - \frac{\sum_{i=1}^{n}\left(X_{pr,i} - X_{ex,i}\right)^2}{\sum_{i=1}^{n}\left(X_{ex,i}\right)^2} \tag{5.52}$$

5.5.4 Sum of Squared Errors

The sum of squared errors (SSE) is the difference between the experimental and predicted values of the model.

$$\text{SSE} = \sum_{i=1}^{n}\left(X_{ex,i} - X_{pr,i}\right)^2 \tag{5.53}$$

5.5.5 Root Mean Square Error (RMSE)

This measures the difference between the predicted value of the model and the experimental data.

$$\text{RMSE} = \sqrt{\frac{\sum_{i=1}^{n}\left(X_{pr,i} - X_{ex,i}\right)^2}{n}} \tag{5.54}$$

5.5.6 Mean Percentage Error (MPE)

The MPE is the average percentage error between predicted values and experimental values.

$$P = \frac{100}{N} \sum_{i=1}^{n} \left| \frac{X_{ex,i} - X_{pr,i}}{X_{ex,i}} \right| \tag{5.55}$$

5.5.7 Mean Square Error (MSE)

The mean square error is one of the statistical tools that quantify the difference between predicted value and actual value. It is the average of the sum squared error. This is calculated as

$$MSE = \frac{\sum_{i=1}^{n} \left(X_{pr,i} - X_{ex,i} \right)^2}{N} \tag{5.56}$$

5.5.8 Adjusted R^2 ($\bar{R}^2$)

The Adjusted R^2 ($\bar{R}^2$) is used to assess the nature and value of R^2 automatically, and errors increase when extra variables are included in the model. However, the values of R^2 and adjusted R^2 will give similar results for a large sample of data.

$$\bar{R}^2 = 1 - \left(1 - R^2 \right) \frac{n_1 - 1}{n_1 - m_1 - 1} \tag{5.57}$$

m_1 and n_1 are the number of repressors in the model and the sample size.

5.5.8.1 Case Study

Chauhan and Kumar (2018) have carried out thermal modelling of a north-wall-insulated greenhouse dryer under both active and passive modes (as shown in Figure 5.4) for the drying of Indian gooseberries (amla).

The gooseberries are small in size, so are dried whole. Therefore they take longer to dry than other agricultural produce. The drying time for goosberries was five days in both dryers. The three most sensitive parameters were selected to be predicted by thermal modeling, namely crop temperature,

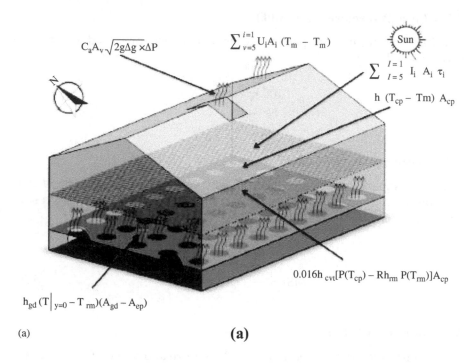

$C_aA_v\sqrt{2g\Delta g}\times\Delta P$

$\sum_{v=5}^{i=1}U_iA_i\,(T_m - T_m)$

Sun

$\sum_{I=5}^{I=1} I_i\,A_i\,\tau_i$

$h\,(T_{cp} - Tm)\,A_{cp}$

$0.016h_{cvt}[P(T_{cp}) - Rh_{rm}\,P(T_{rm})]A_{cp}$

$h_{gd}\,(T\big|_{y=0} - T_{rm})(A_{gd} - A_{ep})$

(a)

(a)

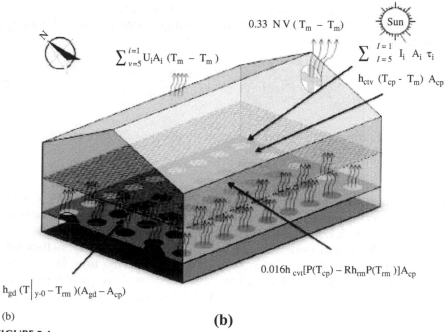

$0.33\,N\,V\,(T_m - T_m)$

Sun

$\sum_{v=5}^{i=1}U_iA_i\,(T_m - T_m)$

$\sum_{I=5}^{I=1} I_i\,A_i\,\tau_i$

$h_{ctv}\,(T_{cp} - T_m)\,A_{cp}$

$0.016h_{cvt}[P(T_{cp}) - Rh_{rm}P(T_{rm})]A_{cp}$

$h_{gd}\,(T\big|_{y-0} - T_{rm})(A_{gd} - A_{cp})$

(b)

(b)

FIGURE 5.4

Schematic diagram of the greenhouse dryer with thermal energy flow for (a) passive mode, and (b) active mode of the insulated north wall greenhouse dryer.

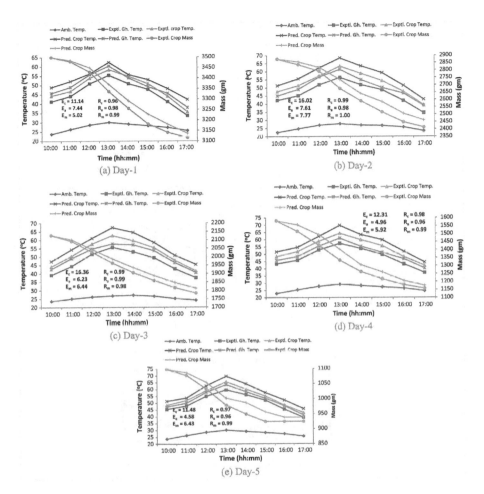

FIGURE 5.5

Experimental observation and predicted outcome for gooseberries under the passive mode of greenhouse dryer.

greenhouse room temperature, and crop weight. Figures 5.5 and 5.6 present the experimental data along with the predicted data on an hourly basis for the whole five days.

To validate the predicted data from the thermal modeling, two main statistical tools were applied in both modes of operation, namely root mean square of percentage deviation, and the coefficient of correlation. A detailed presentation of all the statistical analysis is presented in Table 5.4. The study showed that the predicted and experimental values agree with a high level of accuracy.

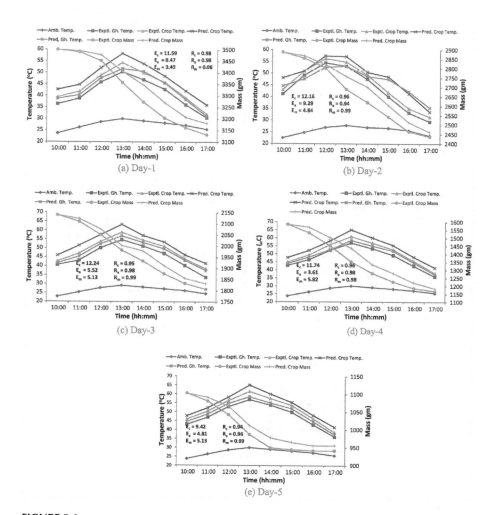

FIGURE 5.6
Experimental observation and predicted outcome for gooseberries under the active mode of greenhouse dryer.

TABLE 5.4

Statistical Analysis of Thermal Modelling of the North-Wall-Insulated Greenhouse Dryer in (a) Passive Mode, and (b) Active Mode

S.No	Drying Day	E_c (%)	E_a (%)	E_m (%)	R_c	R_a	R_m
(a)							
1	1st	11.14	7.44	5.02	0.96	0.98	0.99
2	2nd	16.02	7.61	7.77	0.99	0.99	1.00
3	3rd	16.36	6.23	6.44	0.99	0.99	0.98
4	4th	12.31	4.96	5.92	0.98	0.96	0.99
5	5th	11.48	4.58	6.43	097	0.96	0.99
(b)							
1	1st	11.59	8.47	3.49	0.98	0.98	0.98
2	2nd	12.16	9.29	4.84	0.96	0.94	0.99
3	3rd	12.24	5.52	5.13	0.95	0.98	0.99
4	4th	11.72	3.61	5.82	0.98	0.98	0.98
5	5th	9.42	4.81	5.13	0.97	0.96	0.99

Problems

5.1 What is the significance of convective heat transfer in the drying of crops?

5.2 Derive the expression for the convective heat transfer coefficient under natural and forced modes of drying.

5.3 Discuss the effects of open sun drying and greenhouse drying (under natural and forced convection) on the convective heat transfer coefficient.

5.4 How is thermal modeling helpful for solar dryers?

5.5 Derive an expression for solar temperature under open sun drying for steady state conditions.

5.6 Write the energy balance for various components of greenhouse crop dryers under natural and forced modes of operation.

5.7 Derive an expression for a coefficient of diffusion under a natural convection mode of greenhouse crop dryer.

References

Ahmad, A. and Prakash, O., 2019. Thermal analysis of north wall insulated greenhouse dryer at different bed conditions operating under natural convection mode. *Environmental Progress & Sustainable Energy*, 38(6), e13257.

Ahmad, A. and Prakash, O., 2020. Performance evaluation of a solar greenhouse dryer at different bed conditions under passive mode. *Journal of Solar Energy Engineering*, 142(1).

Arun, K.R., Srinivas, M., Saleel, C.A. and Jayaraj, S., 2019. Active drying of unripened bananas (Musa Nendra) in a multi-tray mixed-mode solar cabinet dryer with backup energy storage. *Solar Energy*, 188, 1002–1012.

Chauhan, P.S. and Kumar, A., 2018. Thermal modeling and drying kinetics of gooseberry drying inside north wall insulated greenhouse dryer. *Applied Thermal Engineering*, 130, 587–597.

Chauhan, P.S., Kumar, A., Nuntadusit, C. and Banout, J., 2018. Thermal modeling and drying kinetics of bitter gourd flakes drying in modified greenhouse dryer. *Renewable Energy*, 118, 799–813.

Kumar, A. and Tiwari, G.N., 2006a. Thermal modelling of a forced convection greenhouse drying system for jaggery: An experimental validation. *International Journal of Agricultural Research*, 1(3), 265–279.

Kumar, A. and Tiwari, G.N., 2006b. Thermal modelling of a forced convection greenhouse drying system for jaggery: An experimental validation. *Solar Energy*, 80, 1135–1144.

Kumar, A. and Tiwari, G.N., 2006c. Effect of shape and size on convective mass transfer coefficient during greenhouse drying (GHD) of Jaggery. *Journal of Food Engineering*, 73, 121–134.

Kumar, A. and Tiwari, G.N., 2007. Effect of mass on convective mass transfer coefficient during open sun and greenhouse drying of onion flakes. *Journal of Food Engineering*, 79, 1337–1350.

Prakash, O. and Kumar, A., 2014a. Design, development, and testing of a modified greenhouse dryer under conditions of natural Convection. *Heat Transfer Research*, 45(5), 433–451.

Prakash, O. and Kumar, A., 2014b. Performance evaluation of greenhouse dryer with opaque north wall. *Heat Mass Transfer*, 50, 493–500.

Qiu, G., Sun, J., Nie, L., Ma, Y., Cai, W. and Shen, C., 2020. Theoretical study on heat transfer characteristics of a finned tube used in the collector/evaporator under solar radiation. *Applied Thermal Engineering*, 165, 114564.

Singh, R.G. and Tiwari, G.N., 2020. Simulation performance of single slope solar still by using iteration method for convective heat transfer coefficient. *Groundwater for Sustainable Development*, 10, 100287.

Tiwari, G.N., 2003. *Greenhouse Technology for Controlled Environment*. Alpha Science, UK.

6

Energy Analysis of Solar Drying Systems

6.1 Introduction

As well as the depletion of fossil fuels in the modern world, the burning of such fuels increases air pollution and leads to global warming, so it is vitally important to deal with these issues (Kumar and Tiwari, 2006; Bagheri et al., 2015; Chauhan and Kumar, 2016; Condori et al., 2001; Jain and Tewari, 2015). Switching to renewable energy is the best solution for these problems, and solar energy is one of the best forms of renewable energy (Reyes et al., 2014; Shalaby and Bek, 2014; Scanlin et al., 1999; Simate, 2001; Xu et al., 2018). One use for solar energy is solar drying, which has long been used in the agriculture sector for the preservation of agricultural produce such as vegetables, etc.

The shortages of fossil fuel mean a reduction in the use of electricity, which became a problem for farmers running agricultural machinery (Shrivastava and Kumar, 2017). Solar dryers help to reduce food loss after harvest. An experiment was conducted on drying fenugreek, putting 2 kg on each tray and taking measurements such as cost analysis, CO_2 emissions, carbon credits, and energy payback time (EPBT). It was found that embodied energy was 1081.83 kWh while EPBT and CO_2 emissions are 4.36 years and 391.52 kg per year.

Solar dryers have always been useful for the betterment of farmers. The hybrid photovoltaic/thermal type of dryer has been developed and tested for drying amla (gooseberries). This consists of a photovoltaic panel (100 wp) and crop dryer (Sajith and Muraleedharan, 2014). The dryer produces both electrical and thermal energy. Also, the performance of the dryer is compared to open sun drying. It was found that the payback time of the dryer was 5.66 years, which is very low compared to the 20 years life span of the dryer.

A solar dryer with phase change material (PCM) has been developed for drying agriculture produce as solar dryers play an imperative role in the conservation of energy (Ananno et al., 2020). An indirect solar heating system has been designed which heats the air by transferring heat from the material being dried. Paraffin is used as the PCM, which absorbs heat during the period of solar radiation and releases it at night, which helps to maintain the continuous process of drying. Economic analysis is carried out using the payback period method, in which payback time, the operational costs of the dryer, and cost of the dried product is determined.

A study of the annual exergy and thermal performance of photovoltaic/ thermal greenhouse dryers has been carried out at IIT Delhi by using various silicon and non-silicon photovoltaic modules such as cadmium telluride, nanocrystalline silicon, crystalline silicon (c-si), high-performance p-type multicrystalline silicon (mc-Si), and iridium gallium selenide (Nayak et al., 2014). The net saving per annum was determined in four types of weather. The study has determined the energy payback time, life cycle, electricity production factor, and life cycle conversion efficiency of the system's embodied energy. It has been found that c-si type photovoltaic modules are best for achieving the maximum carbon credit, energy payback time, and life cycle and CO_2 mitigation.

A double-glazed solar air heater (SAH) was designed, fabricated and tested with paraffin as a PCM, and the experiment was conducted in the climatic conditions of Mashhad, Iran, at latitude 37028IN and longitude 570 20IE for three days during the summer (Edalatpour et al., 2016). The PCM was used to store heat in the form of latent heat, and direct heat in the daytime, and the heat was released at night when there was no solar radiation. Energy and exergy efficiencies were determined by the first and second laws of thermodynamics. It was found that energy and exergy efficiencies were different every day, with energy efficiencies varying from 58.33% to 68.77%, while exergy efficiencies varied between 14.45% and 26.34%. From the economic analysis, it was determined that the cost of heating 1 kg of air for a double glazed solar air heater was US$0.0036.

The main focus was to determine performance analysis. An environmental analysis and mathematical modeling was carried out for drying tomato flakes in a modified greenhouse dryer in an active mode condition (Prakash and Kumar, 2014a). The experiment was conducted in both natural and designed dryers. It was noted that the moisture content varied from 96.0% wb to 9.09% wb in 15 h. The parameters defined by the experiment were embodied energy, CO_2 emissions per year, earned carbon credit, EPBT, and economic analysis. It was found that the payback period of the cost of the dryer was 1.9 years, embodied energy 628.7287 kWh, EPBT was 1.14 years, CO_2 emission/year was 17.6, and net CO_2 emission 38.06 tonnes. The earned carbon credit varied from 12,561.70 INR to 50,245.49 INR during an entire lifespan. Also, it was found that tomatoes dried in dryers contained more nutrients, and the total experiment's uncertainty was 23.41% (Figures 6.1(a) and 6.1(b)).

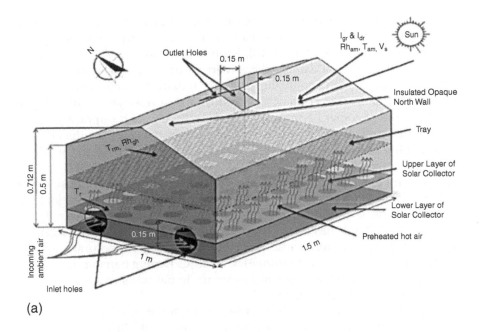

(a)

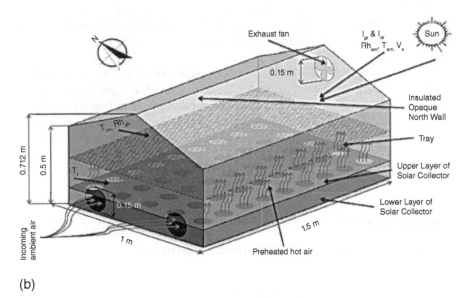

(b)

FIGURE 6.1
(a) Working principle of a natural convection greenhouse dryer (Choab et al., 2019). (b) Working principle of a forced convection greenhouse dryer (Choab et al., 2019).

A hybrid type of solar dryer was designed and fabricated. On sunny days the normal solar dryer was used, while on cloudy days the hybrid dryer replaced it (Amer et al., 2010). Ripe banana slices were dried in the solar dryer. Thirty kg of banana slices were dried for 8 h on sunny days. It was found that in open sun drying, the initial dampness content was 82%, and the final dampness content 62%, but by using the hybrid solar dryer, initial dampness content remained at 82%, but the final dampness content was 18%. It was also observed that in solar dryers the color, aroma and the surface of the product were of better quality than crops dried in the direct sun.

The tentative performance of a photovoltaic ventilated greenhouse dryer used for drying bananas and peeled longans (small, round fruits looking similar to lychees – scientific name *Dimocarpus longan*) has been analyzed (Janjai et al., 2009). It has been observed that for drying peeled longans it takes three days in a greenhouse dryer but 5–6 days with natural sun drying. The air temperatures vary from 31°C to 58°C. Drying banana slices takes four days in a greenhouse dryer and 5–6 days in natural sun drying; the air temperature varies from 30°C to 60°C.

Natural rubber sheets have been dried using the mixed mode and indirect mode solar drying systems (Dejchanchaiwong et al., 2016). It was found that the efficiency of the mixed-mode dryer, i.e. 15.4% more than the indirect mode solar drying system (13.3%) it was also found that mixed-mode was the best to compare to an indirect mode solar dryer; hence it is suggested for smallholders.

Several studies of diverse kinds of greenhouse dryer have been carried out. The choice of a greenhouse dryer for drying a specific crop is the key test. The objective of this review is to offer comprehensive evidence from investigations carried out on greenhouse drying systems, as well as to benefit the inventors, scholars and designers who are developing suitable greenhouse dryers.

6.2 Embodied Energy of Different Prominent Materials

The entire amount of energy used to make any product or to complete any action is known as the embodied energy (Singh et al., 2018). The embodied energy of selected materials is presented in Table 6.1. The embodied energy is presented as either kWh/kg or kJ/kg. Both units are popular among practitioners. Here, 1 kWh is 3.6 MJ.

TABLE 6.1

Embodied Energy of Selected Materials

S.No.	Material	Embodied Energy	Reference
1	Aluminium	55.28 (kWh\kg)	(Baird and Haslam, 1997)
2	Steel	8.89 (kWh\kg)	(Baird and Haslam, 1997)
3	Galvanized iron	9.67 (kWh\kg)	(Baird and Haslam, 1997)
4	Copper	19.61 (kWh\kg)	(Baird and Haslam, 1997)
5	Glass	7.28 (kWh\kg)	(Baird and Haslam, 1997)
6	Polyvinyl chloride	19.44 (kWh\kg)	(Baird and Haslam, 1997)
7	Wood	0.31 (kWh\kg)	(Baird and Haslam, 1997)
8	Wood board	2.89 (kWh\kg)	(Baird and Haslam, 1997)
9	Black paint	25.11 (kWh\kg)	(Baird and Haslam, 1997)
10	Polycrystalline cell	1130.5600 (kWh/m²)	(Prakash et al., 2016a)
11	Battery	46.0000 (kWh/m²)	(Prakash et al., 2016a)
12	Solar charge controller	33.0000 (kWh/m²)	(Prakash et al., 2016b)
13	Polycarbonate	10.1974 (kWh/kg)	(Prakash et al., 2016b)
14	Rubber gasket	25.64 (kWh/kg)	(Nawaza and Tiwari, 2006)
15	Silver coating	0.2780 (kWh/m²)	(Prakash et al., 2016a)

6.3 Energy Payback Time

The energy payback time for a system is the time span needed to recover the embodied energy of the system. Here the system relates to the product and the process. This concept is applicable to both. It is similar to the economic payback time, the only difference being that here emphasis is on energy rather than money (Prakash and Kumar, 2014b). It is expressed in years. Mathematically it is evaluated as follows.

It can be calculated as:

$$E_{\text{payback time}} = \frac{E_{embodied}}{\text{Energy output per year}} \qquad (6.1)$$

6.4 CO_2 Emissions

Watt et al. (1998) observed that during the generation of electricity from coal, 1 kWh of power generation emits 0.98 kg of CO_2.

Hence, CO_2 emissions per year can be expressed as follows:

$$CO_2 \text{ emissions per year} = \frac{E_{embodied} \times 0.98}{L.T.} \qquad (6.2)$$

Where L.T. is the life span of the system. In general, PV panels have around a 35-year life span.

6.5 Earned Carbon Credit

The earned carbon credit is equivalent to:

$$\text{Earned carbon credit} = \text{net mitigation of } CO_2 \text{ in life span} (\text{tones}) \times R \qquad (6.3)$$

Where R is the multiplying factor, which varies between US$5 and US$20/ tonnes of CO_2 mitigation.

The net mitigation of CO_2 across the life span (kg) is calculated as follows:

$$\text{Total } CO_2 \text{ mitigation} - \text{Total } CO_2 \text{ emission}$$
$$= \left[(E_{th} \times n_s) \times \frac{1}{1-L_b} \times \frac{1}{1-L_t} \times 0.98 - E_i \right], kg \qquad (6.4)$$

Where E_{th} is the thermal output of the dryer in kWh, n_s is the life span of the dryer (generally taken to be 25 years), E_i is the embodied energy of the dryer in kWh, L_b is the transmission power loss, which is generally 10%, and L_t transmission and distribution loss, which is also 10%.

6.6 Energy Analysis of Various Solar Dryers

B. Bolaji presents an exergetic analysis of different types of solar dryer (Bolaji, 2011). For the direct mode solar dryer, the energy conversion was found to be 78.1%, while for mixed-mode and indirect mode it was found to be 77% and 49.3%, respectively. Hence it has been observed that mixed mode is better than direct mode, and further it is better than the indirect mode of such dryers. Icier et al. have undertaken a comparison of the exergetic and energetic performances of two different types of drying system, the fluid bed dryer and the tray dryer. Parsley was the crop chosen to be dried in the proposed system (Icier et al., 2008), and the fluid bed dryer was found to be more efficient compared to the tray dryer. Akinola and Fapetu have evaluated the exergetic performance of a dryer operating under mixed-mode for the drying of peppers, waterleaf (*Talinum fruticosum*), okra and yam (Akinola and Fapetu, 2006). The overall thermal efficiency and exergy efficiency were found to be 66.9% and 56%, respectively. The wastage of energy was found to be 44%. Finally, it was noted that this type of dryer is more efficient in comparison to the direct mode drying system. Chowdhury et al. have studied the exergetic and energetic performance of tunnel type solar dryers for the drying of jackfruit leather, which reduce the moisture content from 76% to 11.00% on a wet basis in two days of drying (Chowdhury et al., 2011). However, in open sun drying, the reduction of the moisture increases to 13% on a wet basis for the same drying duration. With solar irradiation of between 100 and 600 W/m², the average exergy and energy efficiency of the dryer were found to be 41.4% and 42.5%, respectively. Fudholi et al. designed and fabricated an indirect mode solar dryer operating under forced convection for drying. The red seaweed was used as the crop to dry in the proposed system. The moisture of red seaweed was reduced to 10% from 90% in 15-h, with solar radiation of 500 Watts per sq m at the flow rate of air of 0.05 kg per second (Fudholi et al., 2014). The efficiency of the dryer was found to be 95%, 25%, and 35%, respectively, across three operating conditions. The Page thin layer drying model was found to be suitable for the drying of red seaweed. Akpinar and Sarsilmaz have presented an exergy and energy analysis of a rotary column cylindrical type of dryer for drying apricots (Akpinar and Sarsilmaz, 2004). The drying temperatures varied from 38°C to 57°C, the relative humidity from 20.5% to 49.7%, and air velocity from 2.3 to 2.5 m/s. The rotary speed was about 2.25 rpm. The maximum losses of exergy were reported as 0.5612 kJ/kg over 76 h of drying. Mokhtarian et al. analyzed the drying performance of different solar dryers working on various air recycling methods for the drying of cassia vera (cinnamon), and the reduction in moisture was found to be between 40% and 25% on a wet basis (Mokhtarian et al., 2016).

Hence it has been observed from various literature reviews that the performance of drying systems can be enhanced, and the losses of energy minimised by using certain techniques such as the recirculation of drying air, plane refractors, thermal storage, and changes in the design of collector units (Akbulut and Durmus, 2010; Panwar, 2014). It also depends on parameters such as drying air temperature, humidity ratio, flow rate, and sample thickness.

6.7 Energy Efficiency of Solar Drying Systems

Because of steep rises in the prices of fuel and raw materials, it is important to focus on the issue of energy efficiency (Prakash and Kumar, 2014b). Energy analysis consists of embodied energy analysis, CO_2 emissions per year, energy payback period, mitigation, and carbon credits. Energy analysis also helps us to discover the rate of solar energy received by solar heaters and accessible for the dryer. Energy analysis can be explained based on mass and energy equations in a steady-state condition (Prakash and Kumar, 2013, 2014c; Madrid et al., 2016; Rabha et al., 2017).

The equation for the conservation of mass of drying air in a solar dryer can be expressed as follows:

$$\Sigma m_{adin} = \Sigma m_{adot} \tag{6.5}$$

The equation for the conservation of the moisture of drying air:

$$\Sigma\left(m_{adot}W_{in} + m_{xp}\right) = \Sigma m_{adot}W_{ot} \tag{6.6}$$

Energy conservation equation:

$$Q - W = \Sigma m_{adot}\left(h_{adot} + \frac{V_{adot}^2}{2}\right) - \Sigma m_{adin}\left(h_{adin} + \frac{V_{adin}^2}{2}\right) \tag{6.7}$$

Energy received by the collector is determined by:

$$Q_{cad} = m_{ad}C_{pad}\left(T_{acod} - T_{acin}\right) \tag{6.8}$$

Inlet and outlet conditions are determined by:

$$\begin{aligned} T_{adin} &= T_{a\,cot} \\ W_{adin} &= W_{a\,cot} \\ \phi_{adin} &= \phi_{a\,cot} \\ h_{adin} &= h_{a\,cot} \end{aligned} \tag{6.9}$$

$$W_{adot} = W_{adin} + \frac{m_{xp}}{m_{ad}} \tag{6.10}$$

Energy used:

$$Q_{ad} = m_{ad}\left(h_{adin} - h_{adot}\right) \tag{6.11}$$

Some exceptions are made for writing energy balance equations, such as neglecting the heat capacity of the greenhouse wall material and cover; air temperature of the greenhouse is not affected by stratification; and the heat capacity of the enclosed air is negligible (Prakash and Kumar, 2015; Ahmad and Prakash, 2019; Ahmad and Prakash, 2020).

6.7.1 Case Study 1: Direct Solar Dryer

Modified types of greenhouse dryer run in both active and passive modes to analyze the energy, exergy, environomical, and annual performance of the dryer (Prakash et al., 2016a). The concept of thermal storage has been applied to the modified type of dryer, with three different modifications: the floor covered with PVC sheets, a bare floor, or a black-coated floor, as shown in Figure 6.2.

The floor of the dryer is made of concrete, which is further covered with black PVC sheeting with hole in the upper layer. The PVC sheet acts as a reflector that resists heat loss to the surface of the ground. An exhaust fan is installed to draw out the humid air, which results in an increase in ground temperature and a decrease in the relative humidity compared to ambient temperature—the rate of change in the moisture of the crop—in this case, tomato flakes, in open type and modified greenhouse dryers. The initial moisture content of the crop in both types of dryer was 96.0% w.b. It was noted that after 10 h of drying in active mode, the mass of the crop (154 g) remained in the dryer, while in the natural mode the mass of the

FIGURE 6.2
Photographs of modified greenhouse dryer under (a) passive mode, and (b) active mode for the drying of potato chips (Prakash et al., 2016b).

crop (169 g) remained in the dryer after 14 h. It was also observed that in the natural mode of the dryer, the amount of crop moisture that was evaporated was equal to 369.5 g, and as the moisture evaporated, the drying rate also decreased. In crops with a high moisture content, such as tomatoes and capsicums, the efficiency of the active mode dryer was higher than that of the passive mode dryer. In contrast, for average moisture content crops such as potato chips, the efficiency of both dryers was equal. It was also observed that crops treated in greenhouse dryers contained more nutrients compared to those dried by using the open drying mode. The payback time for the active mode dryer was 1.89 year, CO_2 emissions per annum were 17.6 kg, and embodied energy was 628.73 kWh, as shown in Tables 6.2 and 6.3. In comparison, for passive mode dryers, the payback time was 1.11 year, CO_2 emissions per annum were 13.4 kg, and embodied energy was 480.27 kWh.

Exergy analysis was conducted in both modes of operation. Various parameters of exergy analysis were evaluated for both modes; namely average exergy efficiency, average exergy, average exergy loss, inlet energy, and average specific energy consumption (see Table 6.4).

6.7.2 Case Study 2: Indirect Solar Dryer

Vijayan et al. (2020) carried out energy and exergy analysis of an active mode indirect solar dryer. The sensible heat storage concept was applied in this dryer. The system was being tested in the southern part of India (Coimbatore) to dry slices of bitter gourd. There are three main component to the system: the solar collector, drying room, and blower, as shown in Figure 6.3.

The system was tested in different air flow rates, namely 0.0141–0.0872 kg/s, respectively. The range of average exergy efficiency at different flow rates was 0.0141–0.0872 kg/s, respectively, as shown in Figure 6.4.

The embodied energy of the system was found to be 1109.307 kWh, and details are shown in Table 6.5.

Various parameters for energy analysis were computed and found to be quite competitive, such as EPBT at 2.21 year, CO_2 mitigation is 33.52 tonnes, and earned carbon credit between INR 10894 and 43576.

6.7.3 Case Study 3: Mixed Solar Dryer

Eltawil et al. (2018) have developed a hybrid or mixed mode tunnel/greenhouse solar dryer. There are three main component to this system, namely, active greenhouse dryer, photovoltaic panel (PV), and solar air heater, as presented in Figure 6.5.

The system was used to dry peppermint. The whole system uses the mixed mode of heat transfer. The embodied energy of the system is presented in Table 6.6.

TABLE 6.2

Embodied Energy for the Fabrication of Modified Greenhouse Dryer

S.No.	Material	Quantity	Embodied Energy Coefficient	Total Energy for Active Dryer (kWh)	Total Energy for Passive Dryer (kWh)
1	Polycarbonate sheet	15.6 kg	10.1974 (kWh/kg)	159.0794	159.0794
2	Glass	5.4 kg	7.28 (kWh/kg)	39.312	39.312
3	Silver coating	0.75 m²	0.278 (kWh/m²)	0.2085	0.2085
4	Black PVC sheet	0.325 kg	19.44 (kWh/kg)	6.318	6.318
5	Wire mesh steel tray	0.7 kg	9.67 (kWh/kg)	6.7669	6.7669
6	Aluminum section				
	(i) 1″×1 mm section	3.59 kg	55.28 (kWh/kg)	198.4552	198.4552
	(ii) 4″×1 mm section	0.82 kg	55.28 (kWh/kg)	45.3296	45.3296
	(iii) 1″×3 mm section	0.080 kg	55.28 (kWh/kg)	4.4224	4.4224
7	Fitting				
	(i) Hinges/Kabja	0.2 kg	55.28 (kWh/kg)	11.0560	11.0560
	(ii) Door lock (hook)	0.025 kg	55.28 (kWh/kg)	1.382	1.382
	(iii) Handle	0.1 kg	55.28 (kWh/kg)	5.5280	5.5280
	(iv) Steel screw	0.25 kg	9.67 (kWh/kg)	2.4175	2.4175
8	DC fan				
	(i) Plastic	0.12	19.44 (kWh/kg)	2.3328	
	(ii) Copper wire	0.050	19.61 (kWh/kg)	0.9805	
9	Polycrystalline solar cell	0.059 m²	1130.6 (Kwh/m²)	66.1378	
10	Battery			46.00	
11	Solar charge controller			33.00	
	Embodied Energy (kWh)			628.7287	480.2776

TABLE 6.3

EPBT of the Dryer, Net Carbon Dioxide Mitigation and Earned Carbon Credit in Various Conditions

S.No.	Parameters	Unit	Potato		Capsicum		Tomato	
			Active	Passive	Active	Passive	Active	Passive
1	EPBT	Years	1.51	1.16	1.24	1.01	1.14	0.94
2	CO₂ mitigation	Tonnes	28.65	28.69	35.01	33.36	38.06	35.36
3	Earned carbon credit (max)	INR	37,826.37	37,826.37	46,222.84	43,620.07	50,245.49	11,669.91
4	Earned carbon credit (min)	INR	9456.59	9470.45	11,555.71	10,904.904	12,561.70	46,680.95

TABLE 6.4

Performance of Modified Greenhouse Solar Drying of Crops and Comparative Analysis of Specific Energy Consumption and Input Energy Under Active and Passive Modes

S.No.	Parameters	Unit	Potato		Capsicum		Tomato	
			Active	Passive	Active	Passive	Active	Passive
1	Average exergy efficiency	%	78.0	86.0	62.0	31.0	30.0	29.0
2	Average exergy	kW	0.016	0.018	0.017	0.017	0.0069	0.00445
3	Average exergy loss	kW	0.0063	0.013	0.0073	0.014	0.012	0.016
4	Inlet energy	kWh	4.559	1.55	4.55	33.16	4.55	4.55
5	Average specific energy consumption	kWh/kg	1.92	1.92	2.99	1.58	1.8	2.58

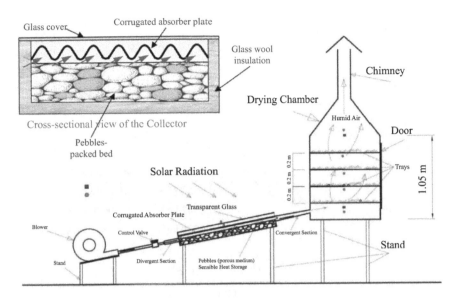

FIGURE 6.3
Schematic diagram of an indirect solar dryer in active mode with sensible heat storage.

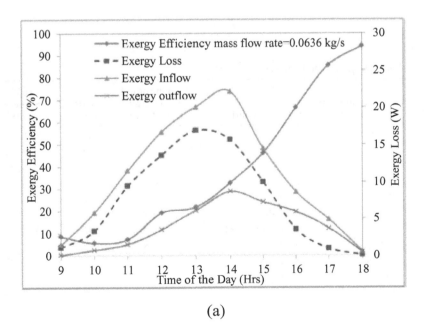

(a)

FIGURE 6.4
(a)–(b) Variation of exergy efficiency and exergy loss during experimentation.   (*Continued*)

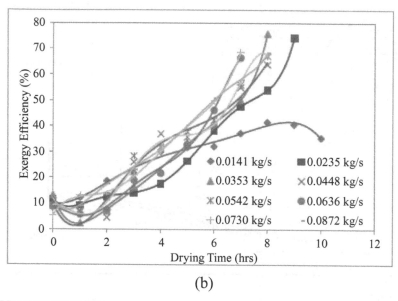

(b)

FIGURE 6.4 (CONTINUED)
(a)–(b) Variation of exergy efficiency and exergy loss during experimentation.

TABLE 6.5

Embodied Energy of the Indirect Solar Dryer

S. No.	Component of Solar Dryer	Material	Embodied Energy (kWh/kg)	Quality (kg)	Total Embodied Energy (kWh)
1	Glass cover	Glass	7.28	4.8	34.944
2	Absorber plate	GI Sheet (0.45 mm thick)	9.636	3.02	29.101
3	Outer cover	GI Sheet (1.2 mm thick)	9.639	14.6	140.686
4	Insulation	Glass wool	4.044	6.45	26.084
5	Piping	Steel	8.89	4.32	38.405
6	Frame	Mild steel	8.89	41.5	368.405
7	Drying trays	Aluminum mesh	55.28	4.54	250.971
8	Fittings	Steel	8.89	1.0	8.890
9	Chimney	PVC pipe	19.39	1.5	29.085
10	Coating	Paint	25.11	2.0	50.220
11	Packed bed	Pebbles	0.0278	110	3.056
12	Blower				
	(i) Copper wire	cooper	19.61	0.5	9.805
	(ii) Casing, fan, shaft, etc.	steel	8.89	13.4	119.126
Total Embodied Energy (kWh)					1109.307

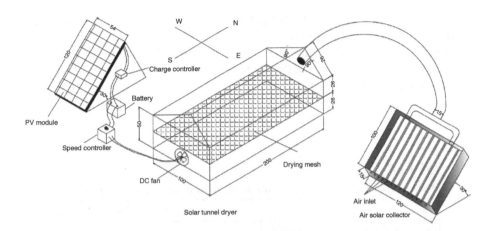

FIGURE 6.5
The schematic diagram of the complete experimental setup (Eltawil et al., 2018).

TABLE 6.6.

Embodied Energy Calculation Data for the Manufacturing of a Mixed-Mode Solar Tunnel Dryer

S.No.	Material	Embodied Energy (kWh/kg)	Weight (kg)	Embodied Energy (kWh)
Collector				
1	Plywood	2.88	7.0	20.16
2	Glass wool	4.044	5.0	20.22
3	Absorber plate	9.639	2.5	24.09
4	PVC pipe	19.39	3.0	58.17
5	Glass cover	7.28	5.5	40.04
6	Paint	25.11	0.5	12.56
Dryer				
1	Polycarbonate sheet	10.1974	30.0	305.922
2	Wire mesh steel tray	9.67	2.0	19.34
3	Hinge	55.28	0.2	11.056
4	Handle	55.28	0.1	5.528
5	Steel screw	9.67	0.2	1.934
6	DC fan			
	1. Plastic	19.44	0.3	5.832
	2. Copper wire	19.61	0.05	0.9805
7	Hooks	55.28	0.1	5.528
Solar cell system				
1	Aluminum section 38×38×3 mm angle	55.28	0.3	16.584
2	Polycrystalline solar cell	1130.60 (kWh/m²)	0.65 m²	734.89
3	Battery			46
4	Solar charge controller			33
Total Embodied Energy (kWh)				**1361.8345**

The system was tested in both load and no-load conditions. Performance details are presented in Table 6.7.

Performance analysis of the system is presented in Table 6.8.

The EPBT of the system is 2.06 years, and net CO_2 mitigation over its life span is 31.80 tonnes.

TABLE 6.7

Daily Average Ambient Conditions and PV Module Energy Output

Item	Test without Load	Test with Load
Insulation of horizontal surface (W/m²)	680	570
Insulation of PV module surface (W/m²)	820	690
Ambient temperature (°C)	24.0	37.0
Output collector temperature	53.6	57.2
Temperature in dryer (°C)	46.45	50.5
Temperature of PV panel (°C)	37.0	55.0
Ambient RH (%)	47.76	33.0
RH at outlet of the collector (%)	7.98	8.87
RH in the tunnel dryer (%)	14.13	23.85
Wind speed (m/s)	1.6	0.5
Energy output of PV module (kWh/day)	0.803	0.796
Energy required to operate DC fan	0.353	0.365
Open circuit voltage (V)	21.03	19.68
Load voltage (V)	15.89	14.02
Load current (A)	2.47	2.9
Short circuit current (A)	4.24	4.49

TABLE 6.8

Calculated Thermal, Drying and Overall Efficiency of the Developed Tunnel Drying System with Different Thicknesses of Layers of Peppermint

Load Density	Collector			Dryer			Overall Efficiency (%)
	Energy Input (MJ)	Energy Output (MJ)	Energy Efficiency (%)	Energy Input (MJ)	Energy Output (MJ)	Energy Efficiency (%)	
I	13.0	8.7	66.91	15.25	4.52	29.62	15.59
II	16.14	9.79	60.66	19.12	5.87	30.71	16.32
III	24.11	10.88	45.12	28.51	7.9	27.72	14.82

Problems

6.1 Discuss the importance of energy analysis.

6.2 Describe some other case studies of different solar drying systems.

Nomenclature

ad	air dryer
ac	solar collector
od	outlet
in	inlet
N	no. of sets
D_{cost}	dryer cost
i	rate of interest
f	rate of inflation
L. T.	life time
Q	net heat rate, kJ/s
m	mass flow rate, kg/s
W	energy utilization rate, kJ/s
V	velocity, m/s
T	temperature, k
ϕ	relative humidity %
h	enthalpy, kJ/kg
o	ambience
C_p	sp. heat at constant pressure, kJ/kgk

References

Ahmad, A. and Prakash, O., 2019. Thermal analysis of north wall insulated greenhouse dryer at different bed conditions operating under natural convection mode. *Environmental Progress & Sustainable Energy*, 38(6), 1–12.

Ahmad, A. and Prakash, O., 2020. Performance evaluation of a solar greenhouse dryer at different bed conditions under passive mode. *Journal of Solar Energy Engineering*, 142(1), 1–10.

Akbulut, A. and Durmus, A., 2010. Energy and exergy analyses of thin layer drying of mulberry in a forced solar dryer. *Energy*, 35(4), 1754–1763.

Akinola, A.O. and Fapetu, O.P., 2006. Exergetic analysis of a mixed mode solar dryer. *Journal of Engineering Applied Science*, 1(3), 205–210.

Akpinar, E.K. and Sarsilmaz, C., 2004. Energy and exergy analyses of drying of apricots in a rotary solar dryer. *International Journal of Exergy*, 1(4), 457–474.

Amer, B.M.A., Hossain, M.A. and Gottschalk, K., 2010. Design and performance evaluation of a new hybrid solar dryer for banana. *Energy Conversation Managament*, 51(4), 813–820, .

Ananno, A.A., Masud, M.H., Dabnichki, P. and Ahmed, A., 2020. Design and numerical analysis of a hybrid geothermal PCM flat plate solar collector dryer for developing countries. *Solar Energy*, 196, 270–286.

Bagheri, H., Arabhoseini, A. and Kianmehr, M.H., 2015. Energy and exergy analyses of thin layer drying of tomato in a forced solar dryer. *Iranian Journal of Biosystems Engineering (Iranian Journal of Agricultural Sciences)* Spring–Summer, 46(1), 39–45.

Baird, A.G. and Haslam, P., 1997. The energy embodied in building materials-updated new zealand coefficients and their significance. *IPENZ Transactions*, 24, 46–54.

Bolaji, B., 2011. Exergetic analysis of solar drying systems. *Natural Resources*, 2(2), 92–97.

Chauhan, P.S. and Kumar, A., 2016. Performance analysis of greenhouse dryer by using insulated north-wall under natural convection mode. *Energy Reports*, 2, 107–116.

Chowdhury, M.M.I., Bala, B.K. and Haque, M.A., 2011. Energy and exergy analysis of the solar drying of jackfruit leather. *Biosystem Engineering*, 110(2), 222–229.

Choab, N., Allouhi, A., El Maakoul, A., Kousksou, T., Saadeddine, S. and Jamil, A., 2019. Review on greenhouse microclimate and application: Design parameters, thermal modeling and simulation, climate controlling technologies. *Solar Energy*, 191, 109–137.

Condorı, M., Echazu, R. and Saravia, L., 2001. Solar drying of sweet pepper and garlic using the tunnel greenhouse drier. *Renewable Energy*, 22(4), 447–460.

Dejchanchaiwong, R., Arkasuwan, A., Kumar, A. and Tekasakul, P., 2016. Mathematical modeling and performance investigation of mixed-mode and indirect solar dryers for natural rubber sheet drying. *Energy for Sustainable Development*, 34, 44–53.

Edalatpour, M., Kianifar, A., Aryana, K. and Tiwari, G.N., 2016. Energy, exergy, and cost analyses of a double-glazed solar air heater using phase change material. *Journal of Renewable and Sustainable Energy*, 8(1), 015101.

Eltawil, M.A., Azam, M.M. and Alghannam, A.O., 2018. Energy analysis of hybrid solar tunnel dryer with PV system and solar collector for drying mint (MenthaViridis). *Journal of Cleaner Production*, 181, 352–364.

Fudholi, A., Sopian, K., Othman, M.Y. and Ruslan, M.H., 2014. Energy and exergy analyses of solar drying system of red seaweed. *Energy and Buildings*, 68, 121–129.

İçier, F., Çolak, N., Erbay, Z., Hancioğlu, E. and Hepbasli, A., 2008. A comparative study on exergetic efficiencies of two different drying processes. *Tarım Makinaları Bilimi Dergisi*, 4(3), 279–284.

Jain, D. and Tewari, P., 2015. Performance of indirect through pass natural convective solar crop dryer with phase change thermal energy storage. *Renewable Energy*, 80, 244–250.

Janjai, S., Lamlert, N., Intawee, P., Mahayothee, B., Bala, B.K., Nagle, M. and Müller, J., 2009. Experimental and simulated performance of a PV-ventilated solar greenhouse dryer for drying of peeled longan and banana. *Solar Energy*, 83(9), 1550–1565.

Kumar, A. and Tiwari, G.N., 2006. Thermal modeling of a natural convection green-house drying system for jaggery: An experimental validation. *Solar Energy*, 80, 1135–1144.

La Madrid, R., Marcelo, D., Orbegoso, E.M. and Saavedra, R., 2016. Heat transfer study on open heat exchangers used in jaggery production modules–Computational Fluid Dynamics simulation and field data assessment. *Energy Conversion and Management*, 125, 107–120.

Mokhtarian, M., Tavakolipour, H. and Kalbasi-Ashtari, A., 2016. Energy and exergy analysis in solar drying of pistachio with air recycling system. *Drying Technology*, 34(12), 1484–1500.

Nayak, S., Kumar, A., Singh, A.K. and Tiwari, G.N., 2014. Energy matrices analysis of hybrid PVT greenhouse dryer by considering various silicon and non-silicon PV modules. *International Journal of Sustainable Energy*, 33(2), 336–348.

Nawaza, I. and Tiwari, G.N., 2006. Embodied energy analysis of photovoltaic (PV) system based on macro- and micro-level. *Energy Policy*, 34, 3144–3152.

Panwar, N.L., 2014. Experimental investigation on energy and exergy analysis of cori-ander (*Coriadrum sativum* L.) leaves drying in natural convection solar dryer. *Applied Solar Energy*, 50(3), 133–137.

Prakash, O. and Kumar, A., 2013. ANFIS prediction model of a modified active green-house dryer in no-load conditions in the month of January. *International Journal of Advanced Computer Research*, 3(1), 220.

Prakash, O. and Kumar, A., 2014a. Environomical analysis and mathematical mod-elling for tomato flakes drying in a modified greenhouse dryer under active mode. *International Journal of Food Engineering*, 10(4), 669–681.

Prakash, O., Kumar, A. and Laguri, V., 2016a. Performance of modified greenhouse dryer with thermal energy storage. *Energy Reports*, 2, 155–162.

Prakash, O. and Kumar, A., 2014b. ANFIS modelling of a natural convection green-house drying system for jaggery: An experimental validation. *International Journal of Sustainable Energy*, 33(2), 316–335.

Prakash, O. and Kumar, A., 2014c. Thermal performance evaluation of modified active greenhouse dryer. *Journal of Building Physics*, 37(4), 395–402.

Prakash, O. and Kumar, A., 2015. Annual performance of a modified greenhouse dryer under passive mode in no-load conditions. *International Journal of Green Energy*, 12(11), 1091–1099.

Prakash, O., Kumar, A. and Laguri V., 2016b. Performance of modified greenhouse dryer with thermal energy storage. *Energy Reports* 2, 155–162

Rabha, D.K., Muthukumar, P. and Somayaji, C., 2017. Energy and exergy analyses of the solar drying processes of ghost chilli pepper and ginger. *Renewable Energy*, 105, 764–773.

Reyes, A., Mahn, A. and Vásquez, F., 2014. Mushrooms dehydration in a hybrid-solar dryer, using a phase change material. *Energy Conversion and Management*, 83, 241–248.

Sajith, K.G. and Muraleedharan, C., 2014. Economic analysis of a hybrid photovoltaic/ thermal solar dryer for drying amla. *International Journal of Engineering Research & Technology (IJERT)*, 3(8), 907–910.

Scanlin, D., Renner, M., Domermuth, D. and Moody, H., 1999. Improving solar food dryers. *Home Power*, 69, 24–34.

Shalaby, S.M. and Bek, M.A., 2014. Experimental investigation of a novel indirect solar dryer implementing PCM as energy storage medium. *Energy Conversion and Management*, 83, 1–8.

Shrivastava, V. and Kumar, A., 2017. Embodied energy analysis of the indirect solar drying unit. *International Journal of Ambient Energy*, 38(3), 280–285.

Simate, I.N., 2001. Simulation of the mixed-mode natural-convection solar drying of maize. *Drying Technology*, 19(6), 1137–1155.

Singh, P., Shrivastava, V. and Kumar, A., 2018. Recent developments in greenhouse solar drying: a review. *Renewable and Sustainable Energy Reviews*, 82, 3250–3262.

Vijayan, S., Arjunan, T.V. and Kumar, A., 2020. Exergo-environmental analysis of an indirect forced convection solar dryer for drying bitter gourd slices. *Renewable Energy*, 146, 2210–2223.

Watt, M., Johnson, A., Ellis M. and Quthred, N. (1998). Life cycle air emission from PV power system. *Progress in Photovoltaic Research Application*, 6(2): 127–136.

Xu, C., Li, X., Xu, G., Xin, T., Yang, Y., Liu, W. and Wang, M., 2018. Energy, exergy and economic analyses of a novel solar-lignite hybrid power generation process using lignite pre-drying. *Energy Conversion and Management*, 170, 19–33.

7

Economic Analysis of Solar Drying Systems

7.1 Introduction

After developing effective solar dryers based on the requirements of potential users, it is very important to also make them effective with respect to their economics, so they can readily be used by farmers and small-scale industries (Aravindh and Sreekumar, 2015). To carry out economic analysis, these are the important considerations:

1. The initial cost of the development of the designed dryer along with all required accessories.
2. Operational cost of the dryer.
3. Annual maintenance cost.
4. Annual cost of the agricultural produce to be dried in the dryer.
5. Life span of the dryer.
6. Salvage value of the dryer.

Over time, researchers have developed various procedures to evaluate solar dryers from the economic point of view. The most important will be discussed in the following sections.

7.2 Cost Analysis

The cost analysis of solar dryers is divided into three sub-sections, namely capital recovery factor, uniform annualized cost, and sinking fund method. Each is discussed below.

7.2.1 Capital Recovery Factor (CRF)

The capital recovery factor (CRF) is the ratio of a constant annuity received from the solar dryer per the life span of the dryer.

Mathematically, it is expressed as

$$CRF = \frac{i(1+i)^n}{(1+i)^n - 1} \qquad (7.1)$$

Where i is interest rate and n is the life span of the solar dryer (years).

7.2.2 Uniform Annualized Cost (UAC)

This is also known as the equivalent annual annuity (EAA). It is the total cost of the solar dryer per year over its life span.

Mathematically, it is expressed as

$$UAC = \frac{P(1+i)^n}{(1+i)^n - 1} \qquad (7.2)$$

Where P is the initial investment in the development of the solar dryer.

7.2.3 Sinking Fund Method (SFM)

The sinking fund method is the technique for depreciating a solar dryer while generating sufficient revenue to replace it at the end of its life span.

Mathematically, it is expressed as

$$SFM = \frac{i}{(1+i)^n - 1} \qquad (7.3)$$

7.3 Cash Flow

Cash flow is considered to be the most important economic concern regarding a solar dryer. In the simplest form, cash flow is the moving in and out of cash within a business.

$$\text{Net cash flow} = \text{Receipts}(\text{Credits}) - \text{Expenses}(\text{Debits}) \qquad (7.4)$$

7.4 Payback Time

Payback time is the time required to get back the amount invested in the solar dryer, with a consideration of interest rates as well as the inflation rate (Shrivastava and Kumar, 2017).

It can be calculated as

$$N = \frac{\ln\left(1 - \frac{D_{cost}}{S}(i-f)\right)}{\ln\left(\frac{1+f}{1+i}\right)} \tag{7.4}$$

Where D_{cost} is the initial cost of the fabrication of the solar dryer, S is the salvage value of the dryer, f is the inflation rate, and i is the interest rate.

7.5 Benefit–Cost (B/C) Ratio

Benefit–cost ratio is the methodology used to select an appropriate dryer, based on considering advantages and disadvantages (Fudholi et al., 2015). It is denoted by the B/C ratio.

The traditionally it is expressed as follow

$$B/C = (\text{Benefits} - \text{Disbenefits})/\text{cost} = (B - D)/C \tag{7.5}$$

Where B is the benefits.

D is the disbenefits.

C is the initial investment.

In the mathematical expression there is no consideration of the operational and maintenance (O&M) costs. Hence it is improved and incorporated the concept below.

Thus the modified B/C ratio is

$$B/C = \frac{B - D - O\&M}{C}$$

A dryer is accepted as suitable provided B/C > 1, and the dryer is rejected if B/C < 1.

7.5.1 Advantages and Limitations of the B/C Ratio

The limitations and advantages are as follows.

(a) It contrasts alternate options over a mutual scale, and enables the testing of different-sized alternate options. It may be used to decide on the best option if it is calculated in increments from the price dimensions.

(b) It could be used to rate the most worthy option for an investor. The benefit/cost ratio system delivers the next advantage, along with other aspects of assessing diverse choices.

7.5.2 Shortcomings of the B/C Ratio

The straightforward benefit/cost ratio cannot be used to decide on the efficiency of a particular dryer. Examination of the benefit/cost ratio is affected by choice concerning the category of the merchandise. It can be an arbitrary choice but has the potential to lead to the analysis of expenditure options.

7.6 Effect of Depreciation

Depreciation is expenditure that decreases over time. It continues to decrease until the life span of the object (in this case, the dryer) is completed. It is represented by D_d.

Mathematically, it is expressed as follows

$$D_d = C - S$$

Where C is the initial cost
  S is the salvage value

Depreciation rate
Depreciation rate is the rate by which the value of the dryer is reduced each year. It is denoted by D_t.

Mathematically, it is expressed as follows

$$D_{t=} = \frac{C - S}{n}$$

Where n is the life span of the dryer (Year).

7.7 Annual Cost Method

The annual cost is the total cost including maintenance of the dryer after excluding the salvage value.

Mathematically, it is expressed as

> Annual cost = Annual initial cost + Annual maintenance cost − Annual salvage value.

7.8 Economic Analysis of Various Solar Dryers

7.8.1 Direct Solar Dryers

Banout et al. (2011) designed and develop an innovative type of direct solar dryer. It was called a double pass solar dryer (DPSD) and is presented in Figure 7.1. The system was tested in load conditions to dry red chilis.

A comparative study is being conduct with the newly developed dryer against a cabinet dryer (CD) and a natural solar dryer or open sun dryer. The comparative experimental values are presented in Table 7.1.

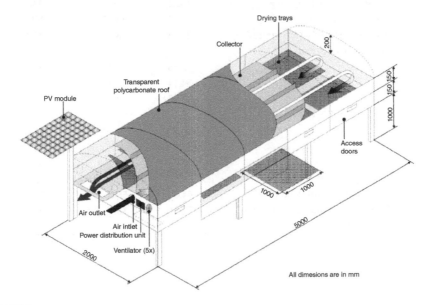

FIGURE 7.1
Schematic diagram of the double pass solar dryer.

TABLE 7.1

Comparative Experimental Value of DPSD with CD and Open Sun Drying

Product Initial Moisture Content (% w.b.)	Red Chili (*Capsicum annum L.*) 90.21		
	Day 1	Day 2	Day 3
Global radiation on the plane of the solar collector (MJ/m^2)	22.20	22.30	23.81
Average ambient temperature (°C)	34.65	33.18	34.17
Average ambient relative humidity (%)	51.89	59.33	57.76
Parameter	DPSD	CD	Open-air sun drying
Quantity loaded (full load) (kg)	38.40	3.30	7.20
Loading density (kg/m^2)	3.66	3.67	3.67
Collector area (m^2)	10.0	0.9	ND
Collector tilt (°)	0	0	0
Solar aperture (m^2)	10.0	0.9	1.96
Tray surface area (m^2)	9.4	0.9	1.96
Air-flow rate (m^3/h)	648.65	64.52	ND
Drying time including nights, up to 10% (w.b.) m.c. (h)	32	73	NOT REACHED
Overall drying efficiency, up to 10% (w.b.) m.c. (%)	24.04	11.52	8.03
First day drying efficiency (%)	15.22	9.32	7.09
Heat collection efficiency (%)	61.62	45.63	ND
Pick-up efficiency, up to 10% (w.b.) m.c. (%)	22.04	18.94	ND
Average temperature of exit air (°C)			
Day 1	52.58	44.45	34.65
Day 2	54.14	45.48	33.18
Day 3	54.52	49.38	34.17
Average relative humidity of exit air (%)			
Day 1	23.65	39.89	51.89
Day 2	24.0	36.70	59.33
Day 3	23.56	31.0	57.76
Maximum drying temp. at no-load (°C)	70.50	60.1	38.80
Maximum drying temp. with load (°C)	64.30	56.0	38.80
Duration of drying air temp. 10 (°C) above ambient temp. (h)	25	18	0

TABLE 7.2

Comparative Economic Evaluation of Double Pass Solar
Dryer and Cabinet Dryer

Parameter	DPSD	CD
Total drier cost (US$)	2700	160
Annual cost (US$)	289.1	32.6
Drying cost (US$/kg)	0.077	0.126
Dried chili obtained per annum (kg)	3724	258
Payback period (years)	3.26	2.42

The economic analysis was compared with a cabinet dryer which is also a direct type solar dryer. Various economic parameters were evaluated, such as total drier cost, annual cost, drying cost per kg of product, and payback period. All costs were evaluated based on recognised international currency, i.e., US dollar ($). The comparative study is presented in Table 7.2.

7.8.2 Mixed Mode Solar Dryers

Elkhadraoui et al. (2015) designed and developed an innovative mixed mode type of solar dryer. In this system an even-span roof type of solar dryer is aided by a solar air heater, so that the system operates in the mixed mode. The system is presented in Figure 7.2.

FIGURE 7.2
Photograph of the mixed mode solar greenhouse dryer.

TABLE 7.3

Economic Parameters for the Mixed Mode Solar Greenhouse Dryer

Capital cost of dryer	2000 DT (Tunisian dinars)
Annual electricity cost of fan	15 DT/year
Capacity of dryer	80 kg (peppers)
	130 kg (grapes)
Price of fresh grapes	1 DT/kg
Price of fresh peppers	0.5 DT/kg
Price of dried grapes	10 DT/kg
Price of dried peppers	10 DT/kg
Life of dryer	20 years
Interest rate	8%
Inflation rate	5%
1 US$= 1.6 D.T.	

An economic analysis was conducted on this system. The data related to this is presented in Table 7.3. The payback period of the system was found to be only 1.6 years.

7.8.3 Indirect Solar Dryers

Sreekumar et al. (2008) designed and developed an innovative indirect solar dryer. This is of the cabinet dryer type. A detail schematic diagram along with all important parts is shown in Figure 7.3, and a photograph of the system in Figure 7.4.

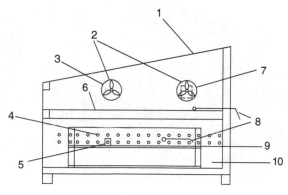

1-Glazing, 2-Fans, 3-Air inlet, 4-Exit of air, 5-Humidity probe, 6-Drier absorber plate,
7-Velocity probe, 8-Temperature sensors, 9-Perforated tray, 10-Drying cabinet

FIGURE 7.3
A schematic diagram of indirect solar dryer.

FIGURE 7.4
Photograph of the indirect solar dryer.

TABLE 7.4

Economic Parameters for the Indirect Solar Dryer

S.No	Parameters	Details
1	Material and labour cost for the construction of the dryer	Rs. 6500.00[a]
2	Interest rate	8%
3	Rate of inflation	5%
4	Real interest rate	3%
5	Life span of the dryer	20 years
6	Electricity cost	Rs. 4.00/kWh
7	Cost of fresh bitter gourd	Rs. 10/kg
8	Selling price of dried bitter gourd	Rs. 250/kg

a US$1 = Rs. 45.

The economic analysis was computed for this system and the important parameters are presented in Table 7.4. The payback period for this system was found to be 3.26 years.

The annualized cost, annual saving, present worth of annual saving and present worth of cumulative saving was also calculated for the whole life time of the dryer. The life span of the dryer is taken as 20 years. There is a detailed presentation of these parameters in Table 7.5. All these costs are taken in INR (Indian Rupees (Rs.)).

TABLE 7.5

A Presentation of the Annualized Cost, Annual Savings, Present Worth of Annual Savings, and Present Worth of Cumulative Savings for the Indirect Solar Dryer

Year	Annualized Cost of Dryer (Rs.)	Annual Savings (Rs.)	Present Worth of Annual Savings (Rs.)	Present Worth of Cumulative Savings (Rs.)
1	920.11	2205.00	2041.66	2041.66
2	920.11	2315.25	1984.95	4026.61
3	920.11	2431.01	1929.81	5956.42
4	920.11	2552.56	1876.20	7832.62
5	920.11	2680.19	1824.06	9656.71
6	920.11	2814.20	1773.42	11430.13
7	920.11	2954.91	1724.16	13154.29
8	920.11	3102.65	1676.26	14830.55
9	920.11	3257.78	1629.70	16460.25
10	920.11	3420.67	1584.43	18044.68
11	920.11	3591.71	1540.42	19585.10
12	920.11	3771.29	1497.63	21082.73
13	920.11	3959.86	1456.03	22538.76
14	920.11	4157.85	1415.58	23954.34
15	920.11	4365.74	1376.26	25330.60
16	920.11	4584.03	1338.03	26668.63
17	920.11	4813.23	1300.86	27969.49
18	920.11	5053.90	1264.73	29234.22
19	920.11	5306.59	1229.6	30463.00
20	920.11	5571.92	1195.44	31659.26

Problems

7.1 Discuss the importance of the economic analysis solar dryers.

7.2 Calculate the payback period of the solar dryer if it was purchased at INR12,000, the interest rate is taken as 6% and inflation rate as 5%, and salvage value is INR2000 after its life span of 20 years.

7.3 Discuss the different kinds of cost analysis.

References

Aravindh, M.A. and Sreekumar, A., 2015. Solar drying—a sustainable way of food processing. In A. Sharma and S. Kumar Kar (eds) *Energy Sustainability through Green Energy* (pp. 27–46). Springer, New Delhi.

Banout, J., Ehl, P., Havlik, J., Lojka, B., Polesny, Z. and Verner, V., 2011. Design and performance evaluation of a Double-pass solar drier for drying of red chilli (*Capsicum annum* L.). *Solar Energy*, 85, 506–515.

Elkhadraoui, A., Kooli, S., Hamdi, I. and Farhat, A., 2015. Experimental investigation and economic evaluation of a new mixedmode solar greenhouse dryer for drying of red pepper and grape. *Renewable Energy*, 77, 1–8.

Fudholi, A., Sopian, K., Bakhtyar, B., Gabbasa, M., Othman, M.Y. and Ruslan, M.H., 2015. Review of solar drying systems with air based solar collectors in Malaysia. *Renewable and Sustainable Energy Reviews*, 51, 1191–1204.

Sreekumar, A., Manikantan, P.E. and Vijayakumar, K.P., 2008. Performance of indirect solar cabinet dryer. *Energy Conversion and Management*, 49, 1388–1395.

Shrivastava, V. and Kumar, A., 2017. Embodied energy analysis of the indirect solar drying unit. *International Journal of Ambient Energy*, 38(3), 280–285.

Appendix

The Microsoft Excel Templates

S.No	Drying Time (Hrs)	Solar Radiation (W/m²)	Ambient Temperature (°C)	Ambient Relative humidity (In decimal)	Wind Speed (m/s)	Inlet Temperature (°C)	Outlet Temperature (°C)	Ground Temperature (°C)	Crop Temperature (°C)	Room Temperature (°C)	Inlet Relative humidity (In decimal)	Outlet Relative humidity (In decimal)	Room Relative humidity (In decimal)	Crop weight (gm)	Moisture Ratio (w.b.)	Moisture Ratio (d.b)	Drying Rate	Important Parameters
																		Tray Area (m²)
																		Ground area (m²)
																		Room Area (m²)
																		Emissivity constant

Index

Milton Keynes UK
Ingram Content Group UK Ltd.
UKHW040051071024
449327UK00019B/489